OPTICAL NETWORKS:
Design and Modelling

IFIP - The International Federation for Information Processing

IFIP was founded in 1960 under the auspices of UNESCO, following the First World Computer Congress held in Paris the previous year. An umbrella organization for societies working in information processing, IFIP's aim is two-fold: to support information processing within its member countries and to encourage technology transfer to developing nations. As its mission statement clearly states,

IFIP's mission is to be the leading, truly international, apolitical organization which encourages and assists in the development, exploitation and application of information technology for the benefit of all people.

IFIP is a non-profitmaking organization, run almost solely by 2500 volunteers. It operates through a number of technical committees, which organize events and publications. IFIP's events range from an international congress to local seminars, but the most important are:

- The IFIP World Computer Congress, held every second year;
- open conferences;
- working conferences.

The flagship event is the IFIP World Computer Congress, at which both invited and contributed papers are presented. Contributed papers are rigorously refereed and the rejection rate is high.

As with the Congress, participation in the open conferences is open to all and papers may be invited or submitted. Again, submitted papers are stringently refereed.

The working conferences are structured differently. They are usually run by a working group and attendance is small and by invitation only. Their purpose is to create an atmosphere conducive to innovation and development. Refereeing is less rigorous and papers are subjected to extensive group discussion.

Publications arising from IFIP events vary. The papers presented at the IFIP World Computer Congress and at open conferences are published as conference proceedings, while the results of the working conferences are often published as collections of selected and edited papers.

Any national society whose primary activity is in information may apply to become a full member of IFIP, although full membership is restricted to one society per country. Full members are entitled to vote at the annual General Assembly, National societies preferring a less committed involvement may apply for associate or corresponding membership. Associate members enjoy the same benefits as full members, but without voting rights. Corresponding members are not represented in IFIP bodies. Affiliated membership is open to non-national societies, and individual and honorary membership schemes are also offered.

OPTICAL NETWORKS: Design and Modelling

IFIP TC6 Second International Working Conference on Optical Network Design and Modelling (ONDM'98)
February 9-11, 1998
Rome, Italy

Edited by

Giancarlo de Marchis
Fondazione Ugo Bordoni
Italy

Roberto Sabella
Ericsson Telecomunicazioni —CoRiTel
Italy

KLUWER ACADEMIC PUBLISHERS
BOSTON / DORDRECHT / LONDON

Distributors for North, Central and South America:
Kluwer Academic Publishers
101 Philip Drive
Assinippi Park
Norwell, Massachusetts 02061 USA
Telephone (781) 871-6600
Fax (781) 871-6528
E-Mail <kluwer@wkap.com>

Distributors for all other countries:
Kluwer Academic Publishers Group
Distribution Centre
Post Office Box 322
3300 AH Dordrecht, THE NETHERLANDS
Telephone 31 78 6392 392
Fax 31 78 6546 474
E-Mail <services@wkap.nl>

Electronic Services <http://www.wkap.nl>

Library of Congress Cataloging-in-Publication Data

International Working Conference on Optical Network Design and Modelling (2nd : 1998 : Rome, Italy)
Optical networks : design and modelling : IFIP TC6 Second International Working Conference on Optical Network Design and Modelling (ONDM '98), February 9-11, 1998, Rome, Italy / edited by Giancarlo de Marchis, Roberto Sabella.
p. cm. — (International Federation for Information Processing ; 19)
Includes bibliographical references.
ISBN 0-412-84790-6
1. Optical communications Congresses. 2. Fiber optics Congresses. 3. Computer networks Congresses. 4. Optoelectronics Congresses. I. De Marchis, Giancarlo. II. Sabella, Roberto. III. International Federation for Information Processing. Technical Committee 6. IV. Title. V. Series: International Federation for Information Processing (Series) ; 19.
TK5103.59.I59 1998
621.382'7—dc21 99-23737
CIP

Printed on acid-free paper.

Printed in the United States of America.

CONTENTS

PREFACE

Recently the term *optical networking* is widely used by Telecom people, to express the capabilities of routing and switching channels directly in the optical domain without opto/electronic conversion. In fact, all-optical networks are those in which the path between the using nodes at the ends remains entirely optical from end to end. Such paths are usually named *lightpaths*. Optoelectronic technology, which enables the realization of optical networks, are approaching functional and economic feasibility.

As a consequence more and more people worldwide are investigating them as possible base upon which networks of the future can be built, both within the wide-area backbone and for the metropolitan and local area distribution facilities.

The motivations for optical networks success basically stay the possibility of satisfying the growing demand for i) bandwidth per user, ii) protocol transparency, iii) higher path reliability, and iv) simplified operation and management. In all these matters, there is evidence that classical time division multiplexing (TDM) approaches realized in electronic circuitry of ever-increasing speed, variety and complexity, are slowly beginning to prove insufficient; that means either that older technology cannot do the required job as cheaply or that it cannot do it at all and the optical approach can.

Among the different approaches toward realizing high-capacity, protocol transparent optical networks, wavelength division multiplexing (WDM) offers the best promise in the near term. In fact, while enabling meaningful capacity enhancements, WDM technique provides the means for novel networks in which the paths routing is directly accomplished in the wavelengths domain. Indeed, WDM networks offer potential advantages, including higher aggregate bandwidth per fiber, new flexibility for automated network management and control, noise immunity, transparency to different data formats and protocols, low bit-error rates, and better network configurability and survivability: all leading to more cost effective networks.

The papers contained in the present book address several hot topics relating to optical network design and modelling.

Optical networking issues as the routing algorithms and wavelength conversion and assignment are treated in the first two parts, respectively. Examples of the application of optical technology for the realization of local and access networks are reported in the papers of the III part. The key topic of transmission of optical signal within an optical transport network is treated in four papers of the IV and V part. New paradigms relating to traffic models are introduced and discussed in part VI. Other advanced topics in optical network, such as the relation between ATM and optical technologies, are reported in the last part.

OPTICAL NETWORK DESIGN AND MODELING II

Organizing Committee
Harmen R. van As, Vienna University of Technology, Vienna, Austria
Admela Jukan, Vienna University of Technology, Vienna, Austria
Giancarlo de Marchis, Fondazione Ugo Bordoni, Rome, Italy
Roberto Sabella, Ericsson Telecomunicazioni - CoRiTeL, Rome, Italy

Scientific Program Committee
Harmen R. van As, Vienna University of Technology, Vienna, Austria
Eugenio Iannone, Pirelli, Milano, Italy
Hubert Jaeger, Swiss Federal Institute of Technology (ETH), Zurich, Switzerland
Marco Listanti, Universita' di Roma "La Sapienza", Rome, Italy
Giancarlo de Marchis, Fondazione Ugo Bordoni, Rome, Italy
Luigi Fratta , Politecnico di Milano, Milan, Italy
Roberto Sabella, Ericsson Telecomunicazioni - CoRiTeL, Rome, Italy
Bernhard Schmauss, Lucent Technolgies, Nuernberg, Germany

ACKNOWLEDGEMENTS

We wish to warmly thank Stefano Binetti for his patient and precious work to support the final edition of this book. We also like to thank Paola Cauli and Rosario Vozzella for their notable assistance in the organization of the Conference.

PART ONE

Routing algorithms for optical networks

Restoration methods for multi-service optical networks

Admela Jukan and Arnold Monitzer

Vienna University of Technology

Institute of Communication Networks

Gusshausstrasse 25/388, A-1040 Vienna, Austria

Tel. +43-1-58801-3993 Fax +43-01-587 05 83

e-mails: admela.jukan@tuwien.ac.at arnold.monitzer@imladris.ikn.tuwien.ac.at

Abstract

In this paper, we apply different restoration strategies for different optical network services in multi-service optical WDM networks according to the user requirements and availability of network resources. For that purpose optical network services provided to the user networks like SDH or ATM are classified according to the quality degree and restorability required. In particular, the restoration means dynamical re-establishing of circuits, which are generally dynamically or statically assigned. The numerical results for networks with a random restorable topology and four classes of service quality degree as well as for service-specific restoration methods are presented and studied on their applicability.

Keywords

Optical network service, WDM networks, QoS-routing, restorability, blocking probability

INTRODUCTION

One of the key parameters of interest for client networks like SDH (Synchronous Digital Hierarchy) or ATM (Asynchronous Transfer Mode) is the service restorability in optical networks. This issue is even more important in case of optical wavelength division multiplexing (WDM) networks where a single fibre is carrying enormous amount of data and a fibre break has an massive impact on carried traffic. The restoration, as discussed here, refers to attempt to re-establish a connection in optical networks, which is being affected by a signal degrade or by the complete loss. Optical network service restorability (ONSR) relates to the efficiency of the restoration method applied. An optical network service (ONS) is defined as any kind of handling of optical signals originating from optical network clients [1]. Like other quality attributes (transmission, access, ...), the ONSR also varies with performance objectives of each single client network. High restorability may be required for clients handling real-time data, while computer communications for other data may tolerate higher connection blocking probabilities or long restoration times. We do not consider any time sensitive attributes of restoration, i.e. we assume that the restoration time is negligible.

Since the Quality-of-Service (QoS)-guarantee is the main issue for integration of optical networks into the existing network infrastructure, the optical network services are here classified according to the quality degree (QoS-guarantee) and restorability required. By using the QoS-routing method to set-up connections, a route capable of providing all necessary QoS-performances for transmission, routing, network management, economic efficiency and restoration is selected for every connection request to the optical layer [1]. We assume that restoration is needed each time the QoS is degraded. The blocking probability is here defined as probability that for a certain service call the required restorability and hence QoS cannot be provided.

RESTORATION METHODS

The original path that is established at the connection set-up is called the *working path.* In case of a fibre cut a network has to find an alternative temporary path, a *restoration path*, for all the affected services of the damaged link. If a restoration path is pre-reserved either exclusively for a connection or by sharing, it is also called a *spare path.*

In this paper, we distinguish between two restoration schemes depending on restoration end-points of a connection. One is the restoration between the working path-terminating nodes of the failed path, *path-restoration* (PR), the other is the restoration between the terminating nodes of a failed link, *link-restoration* (LR).

In both of these methods either *dynamic-* or *static (pre-planned) restoration* techniques can be applied [2]. With *dynamic restoration* techniques the spare entities in the case of link failure are dynamically searched. In this case the complex routing algorithm must be applied. The main benefits of dynamic restoration methods are robustness to multiple link failures and a simple *a posteriori* network expansion. *Static restoration* techniques reserve spare entities for a connection at the moment of a connection set-up. Since no routing is necessary at the moment of failure, this technique is faster than dynamic restoration. However, the network utilisation is diminishing, because of large number of spare resources reserved.

In this paper, for every existing service a specific restoration method for link failures is assigned. *Dynamic Link Restoration (DLR)* establishes spare paths between link-terminating nodes of a failed link. For all affected services using a failed link a connection for restoration must be found. For *Dynamic Path Restoration (DPR)* restoration paths between path-terminating nodes of all affected paths must be found. The number of optical paths to be restored is directly proportional to the traffic on a failed link. *Static LR (SLR)* is based on pre-planned *1:1* spare links that are assigned at the connection set-up. This is supposed to be necessary only for very limited number of services requiring extremely high restorability. Because of the large waste of capacities, *Static PR (SPR)*, based on pre-planned *1:1* spare paths is preferred. It uses the link disjoint paths algorithm, where network graph transforms for path/link-terminating nodes are applied. This algorithm will be explained in more details in the next section.

As a trade-off between fair network utilisation and restorability the methods *Shared Static LR (SSLR)* and *Shared Static PR (SSPR)* can be used. The SSLR-method means that a certain wavelength is assigned as a spare for more than one spare path on a link. If a failure occurs, only one spare path can be established per shared wavelength. The SSPR-methods means that the resources used for a certain spare path should be preferably used when assigning another spare paths.

DEFINITION OF RESTORABILITY

The restoration ratio or restorability of a service S_i is the measure of efficiency of a restoration method. The ***restorability, r_i***, (*i* indicates the service quality degree) is here defined as:

$$r_i = \frac{\text{successfully restored services } S_i}{\text{number of attempts to restore a service } S_i} \qquad (1)$$

For restoration purposes it will be important to find two shortest link-disjoint paths between two nodes. The motivation to find an algorithm that provides link disjoint paths is explained with Figure 1. In Figure 1a, the spare path is completely separated from the working one. However, as furthermore shown in Figure 1b, if at least one link of the affected path use the same duct as a link of the working path, the recovery of link failure between the nodes 3 and 7 is not possible. This comes from the fact that mostly all fibres that use the same duct are damaged in case of a fibre cut. For that purpose, it is advantageous to find link disjoint paths.

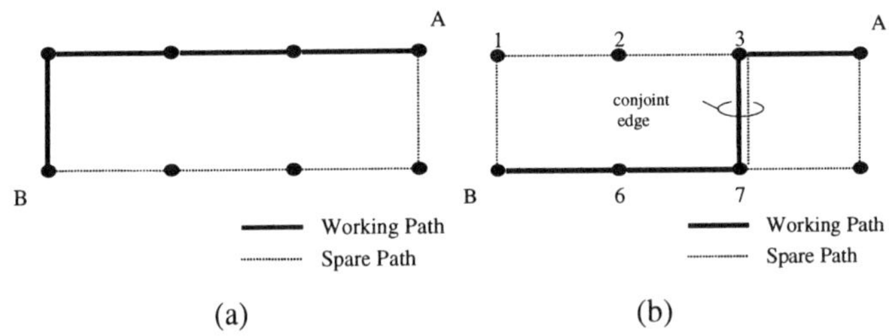

Figure 1 An example where working and spare path are two link disjoint (a) or conjoint (b) paths

It is not always possible to find a link disjoint path for a certain network topology. For that purpose, we first have to test the network topology, if a link disjoint path exists. This is strongly related to the definition of reliable networks. The fundamental requirement for network reliability can be expressed as the condition that any two subparts of the graph must be connected by at least two links [3]. This is the guarantee that in the case of a single link failure the network remains connected and restoration paths can be established along alternative physical paths. For that reason the nodal degree for every node must be greater or equal 2.

Note that the condition must be modified for survivability, if a node-failure occurs. When designing survivable networks we postulate the following:

> *Any two subparts of a graph must be connected by at least two disjoint links, and, if there is more than one node in a sub-graph, at least two links must originate from different nodes.*

For that purpose, not only link disjoint path algorithms, but also node disjoint path algorithms are necessary. According to that, we can say that a bi-directional bus-topology represents a non-survivable graph. On the other side, a bi-directional ring-topology is survivable for link and node failures.

Now the *link disjoint path algorithm (LDPA)* is presented in more detail. Let us assume that for a pre-planned restoration technique it is necessary to find a link-disjoint path for the working path in the network *G (V, E)*, where $V = \{ v_1, v_2, .., v_N \}$ is representing *N* nodes and $E = \{ e_1, e_2, .., e_L \}$ stays for *L* links.

Assume that the algorithm is shown for the path-terminating nodes *src* (source) and *dest* (destination). The routing algorithm *routing(src, dest, G)*, used here is based on routing with QoS, where for every service call a path capable of providing a QoS guarantee is chosen. The services are classified according to the quality degree to be provided for them. As a shortest path algorithm we use the Dijkstra's shortest path algorithm [4].

In Figure 2 it is shown how the link-disjoint path algorithm is accomplished. First a shortest path from node *src* to node *dest*, called $path_1$*=routing(src, dest, G),* is found. The connection $path_1$ between the source and destination node will be established for the connection set-up in a common way. Then in the next step all links of the graph G between the intermediate nodes of the $path_1$ are removed, so that conjoint links are not longer available. Now the graph has to be transformed into a graph *G'*, $G' = G \setminus \{ e_i \}$, $\forall\, e_i \in path_1$. For the transformed path *G'*, the second link disjoint shortest path $path_2$ = *routing(src, dest, G')* must be found. With the last step, the path $path_2$ is finally the link-disjoint path.

The same algorithm can be applied for finding the restoration path for a one link of a working path, so that the path-terminating nodes *src* and *dest* are replaced by the link-terminating nodes.

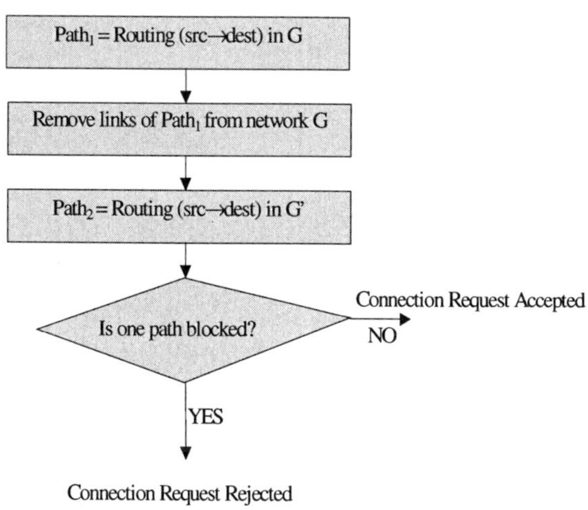

Figure 2 LDP-algorithm

PERFORMANCE STUDY AND SIMULATION RESULTS

For the numerical examples shown in this section, we use the basic assumptions of circuit switching networks:

- Poisson service request arrivals normalised to the mean call duration (a call stands for a connection request),
- exponentially distributed call duration times with unit mean,
- no queuing of calls,
- uniform traffic distribution.

The following assumptions for example network topologies are made:

- network load is defined for the entire network as the ratio between arrival- and duration rates,
- restorable random network topology is studied: randomly generated network for a given number of nodes and connectivity (30 nodes, connectivity 0.31, minimal nodal degree 2),
- equal number of wavelengths is assumed per links for the entire network under consideration,
- user calls for connections of a certain quality are randomly generated,
- all results are obtained with a confidence level of 95% ,
- Dijkstra shortest path algorithm is used.

For QoS-routing we assume the following:

- four different service classes (S_0, S_1, S_2, S_3),
- with 8 wavelengths per link, the following primary wavelength sets are taken: $W_{prim}[S_0]=\{\lambda_1, \lambda_2\}$, $W_{prim}[S_1]=\{\lambda_3, \lambda_4\}$, $W_{prim}[S_2]=\{\lambda_5, \lambda_6\}$, and $W_{prim}[S_3]=\{\lambda_7, \lambda_8\}$,
- wavelengths reserved for lower index (or lower quality) services are generally allowed to use the wavelength of higher-index services (higher quality),
- quality requested from the user side can only be provided by a service of the same or higher quality

Assumptions regarding restoration:

- Poisson process is used as negative exponentially distributed inter-arrival times for link failures (arrival rate is taken as 0.002 time units),
- repair time is deterministic (0 time units, i.e. no any new call is affected by a link failure),
- *LDPA* is used for static restoration methods.

Other parameters will be particularly given.

In Figure 3 a comparison between the degrees of restorability of all introduced methods DPR, DLR, SPR, SLR, SSPR and SSLR in case where no service classification is provided is given. This is shown for a wavelength-shifted network. The SPR and the SLR are both providing a

100% restorability (both graphs coincide), since we are considering only the link failures. For higher network loads, the DPR shows better restorability with respect to DLR. This is because by establishing many paths between the link-terminating nodes, the blocking probability increases, with respect to the case that many different path-terminating nodes have to be established. If the load is kept low, the DLR and DPR are generally providing nearly the same restorability, because the both methods are the dynamic methods, and directly proportional to the probability of connection accommodation (1-blocking probability).

In Figure 4 the restoration methods DPR and SSPR are compared for the networks with and without wavelength shifting. The benefits in blocking probability for wavelength shifted networks become significant with the increased network size [1]. The same is valid for restorability. Higher restorability is provided by the dynamic methods in a wavelength shifted network. However, the results with the method SSPR significantly differ. Here, the networks without wavelength shifting show better restorability. This comes from the wavelength assignment of a wavelength for sharing. For a wavelength continuous network fewer resources are generally shared since the paths are longer, i.e. the probability that a wavelength can be shared is smaller. This is not the case with a wavelength-shifting network.

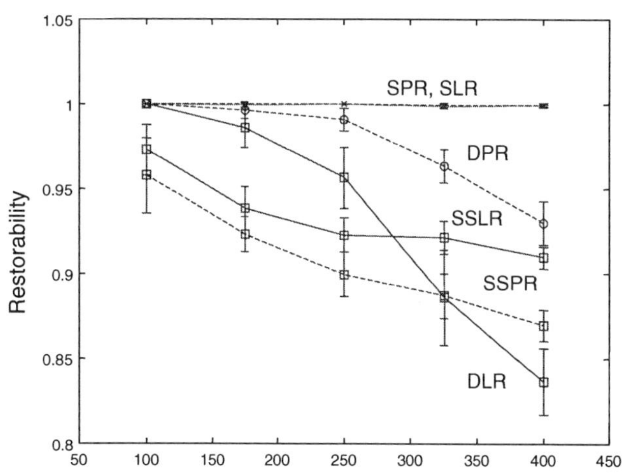

Figure 3 A comparison between different restoration methods for our example network (dashed line- path methods, full line-link methods).

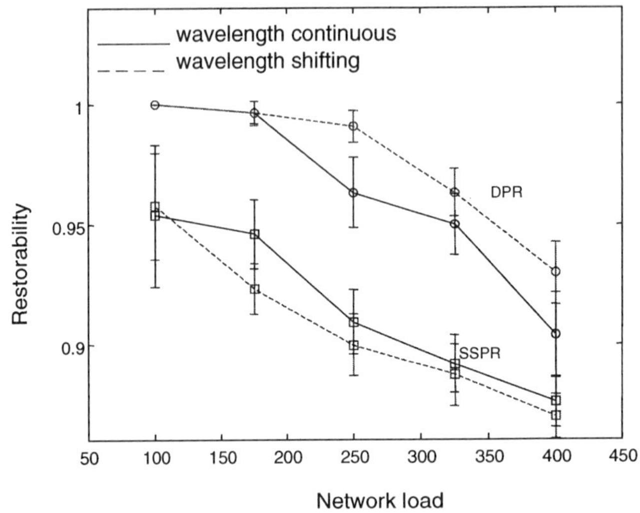

Figure 4 Restorability for DPR and SSPR methods for the example network with and without wavelength shifting.

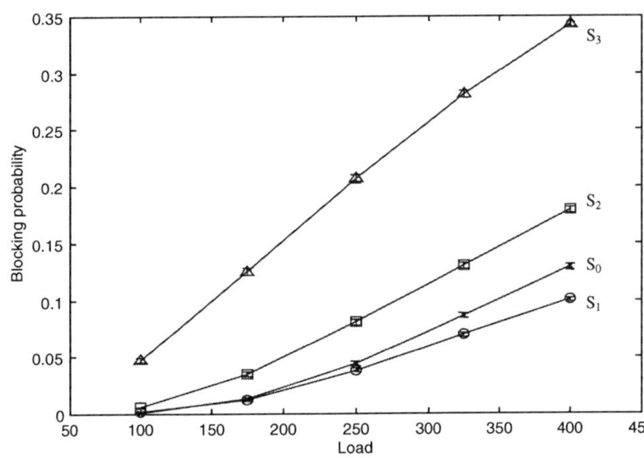

Figure 5 Blocking probability of four services in 30 nodes WDM transport network (wavelength shifted, 100% link failures, (x) S_0 with SPR, (o) S_1 with no restoration, (□), S_2 with DLR, (Λ) service S_3 with DLR)

For the same network we now assume 4 different service classes, where the following restoration methods are assumed: SPR for S_0, no restoration for S_1, DLR for S_2, and DLR S_3. This is shown in Figure 5 and Figure 6. It can be expected that the blocking probability of service S_0 would be the lowest, because all free wavelengths of λ_1-λ_2 are available for this service [1]. Since the SPR method is used for this service, blocking probability increases very much and exceeds blocking probability of service S_1. Blocking probability of services S_1-S_3 will not be changed with respect where no failures happen, because for dynamic methods no spare entities have to be allocated.

In Figure 6, the restorability per service is shown. The service S_0 provides 100% restorability, while the restorability of service S_1 is zero for any network load (for lower network loads, the uncertainty of results increases, since no path will be affected during a sample period of confidence). It can be seen that service S_2 is assigned to a better restorability than service S_3, because more wavelengths can be available to the service S_2.

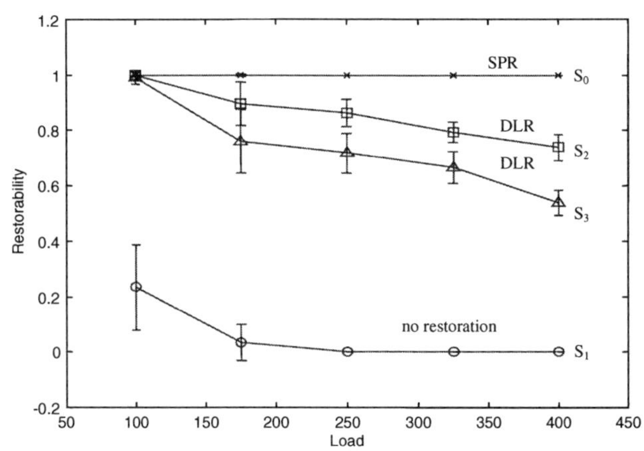

Figure 6 Restorability for different services with different restoration methods assigned (the same assumptions as in Figure 5).

Conclusion

In this work, different restoration methods are applied to different services and connections in optical networks. This comes naturally from the fact that the client networks are having different requirements on the QoS and hence on the restorability for a connection. The restorability and network utilisation efficiency are two opposite conditions to fulfil and a trade-off between service blocking probability and appropriate restorability must be found. We showed that an important issue for the future study is the performance comparison between networks with and without wavelength shifting, which use service-specific restoration methods.

References:

1. A. Jukan, H.R. van As: Quality-of-service routing in optical networks, Proceedings on ECOC'97, Session We2D, pp. 160-163, 1997
2. T.-H. Wu: Fiber Network Survivability, Artech House Inc, Boston-London, 1992.
3. S. Baroni, P. Bayvel: Wavelength Requirements in Arbitrarily Connected Wavelength-Routed Optical Networks, Journal of Lightwave Technology, vol. 15, No.2, Feb. 1997.
4. A. Kershenbaum: Telecommunications Network Design Algorithms, McGraw-Hill, Inc, 1993.

Performance Evaluation of Photonic Networks under Dynamic Traffic Conditions

Jan Späth and Stefan Bodamer
University of Stuttgart
Institute of Communication Networks and Computer Engineering
Pfaffenwaldring 47, D-70569 Stuttgart, Germany
Tel.: +49 711 685 7990, Fax: +49 711 685 7983
E-Mail: {spaeth, bodamer}@ind.uni-stuttgart.de

Abstract

This paper focuses on two aspects of photonic network traffic performance investigation. Two different modelling approaches for the evaluation of network nodes under dynamic traffic conditions are described. This allows to evaluate how suitable different node architectures are for networks carrying dynamic traffic. Furthermore, routing strategies are becoming more important with increasing dynamic. A lot of work reported in literature deals with "wavelength routing". In most cases however, this means static wavelength assignment to routes and not dynamical routing of random wavelength path requests. This paper now considers dynamic traffic aspects of wavelength routing where many analogies to classical routing strategies can be observed. Therefore, a short overview of routing methods is given before we describe our modelling approach and present some case studies and results.

Keywords

Photonic transport network, WDM, performance evaluation, dynamic traffic

1 INTRODUCTION

Photonic technology is able to handle the expected future traffic growth. Although it is not clear how future networks will look in detail, a promising solution is to enhance present networks with a transport layer based on WDM transmission technology and network nodes with cross-connect (CC) functionality. A lot of technological as well as theoretical research has been done. One big field is optimised network planning with the goal of minimising the required network resources. This is usually done for static traffic requirements using more or less complicated optimization strategies. However, real network operation will not only see static traffic requirements but also dynamic behaviour basically due to two reasons: *unexpected network failures* and *random traffic changes*. The latter is caused by the inherent randomness of user behaviour and environmental changes. Therefore the consideration of traffic behaviour plays a central role in all network planning and evaluation methods applied in the past.

In future photonic networks, the dynamic will further increase. Although many details are still not clear and heavily discussed, the following scenario seems to be realistic:

- Today, we see mainly static point-to-point WDM channels, soon supported by protection paths.
- The next step will see wavelength paths passing multiple nodes. First, these paths will be realised via fixed nodes like ADMs, then reconfiguration possibilities will be added including protection schemes.
- Further increase of flexibility could then be achieved by using optical CCs and the possibility for the network management to establish wavelength channels dynamically to adapt to requirement changes.
- In the long-term, at least partly control of channels by user signalling could also be possible.

Since in the foreseeable future the optical layer will serve as a transport layer, the dynamic in this layer is determined by the aggregation of many user traffic streams and therefore not easy to predict. Basically, the dynamic will decrease when going from the user to the transport layer, the granularity however will increase. Therefore, a dynamic consideration of the transport layer is important due to the huge impact of blocking effects in this layer.

2 ROUTING METHODS FOR CIRCUIT-SWITCHED PHOTONIC NETWORKS

Photonic networks will be based on circuit-switching (CS) in the foreseeable future. Various routing methods exist for CS networks and also several classification schemes can be applied. Some classification criteria applicable for CS as well as for packet switched (PS) networks are listed below. An overview of these and more routing strategies can be found in [1].

- **adaptive / non-adaptive**: Adaptive routing methods consider the network situation (traffic load, equipment status, ...) at the moment of a call request whereas for non-adaptive methods routing decisions are made independent of the present network situation.
- **source routing / hop-by-hop routing**: When using source routing schemes, the route is determined at the source node of a call request. If multiple alternatives are available, this node is also responsible for the choice between them. Hop-by-hop routing uses step-wise routing by forwarding the routing control from node to node until the end node is found. The great variety of rules for the forwarding process leads to many different routing strategies.
- **central / distributed / isolated**: All central schemes use a central network control instance (often called Routing Control Centre, RCC) for the routing decisions whereas for isolated schemes each network node makes the decisions based on its own information base without communicating with other nodes. Distributed schemes make use of active information exchange between several nodes.
- **hierarchical / non-hierarchical**: Hierarchical routing looks for routes using the network hierarchy as guideline (e. g. routing in classical telephone networks) whereas non-hierarchical strategies use paths independent of hierarchical properties of the network topology (an example is routing in the internet).

For routing in photonic networks, many analogies exist to classical networks. By comparing the wavelength domain of a photonic network with the time domain of a classical network, one can state the following. From a functional point of view, photonic networks which are fully equipped with wavelength converters can be compared with classical TDM networks with time-slot interchange (fully equipped means, that in every network node each incoming wavelength can be translated to any outgoing wavelength, this is often called the *Virtual Wavelength Path* or VWP concept). Without wavelength conversion (often called *Wavelength Path* or WP concept), the functionality can be compared to TDM networks without time-slot interchange (the time slot respective the wavelength of a connection is fixed throughout the network). Any intermediate wavelength conversion concept results in internal blocking in the nodes.

However, photonic networks provide also several particularities compared to classical networks. The exploitation of the wavelength (or frequency) domain provides a new degree of freedom for routing decisions. This leads to significant enhancement of routing possibilities: in every node a decision is made which link, fibre, and wavelength is chosen for the next section. On the other hand, the analog nature of optical networks results in new restrictions concerning for example maximum path lengths or a maximum number of nodes which can be crossed before electrical signal regeneration is required. Furthermore, the impact of blocking in the optical domain is much higher than in higher network layers due to the huge transport streams carried in the optical domain.

3 MODELLING AND TOOL REALISATION

An important aspect for photonic network evaluation under dynamic traffic is how to model the random behaviour of wavelength channel setups/releases. We use two different modelling approaches depending on whether the focus is on the node architecture or on the mechanisms concerning the network as a whole.

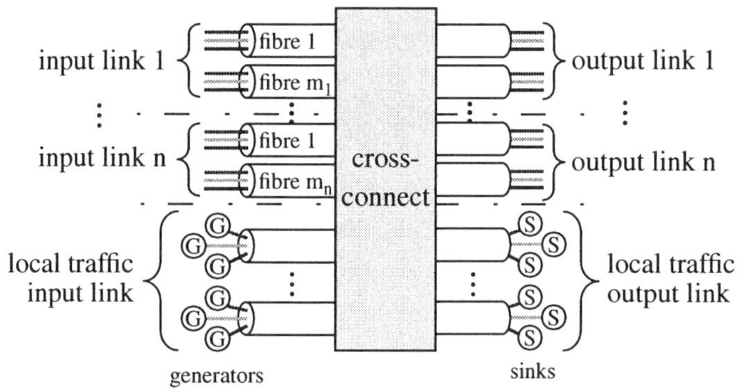

Figure 1 Modelling approach for the evaluation of node architectures

The approach that seems to be appropriate when investigating and comparing different node architectures under dynamic load scenarios uses one traffic generator for each wavelength (see Figure 1). In this case, for each source (i.e. a wavelength channel that is originating at a node) one generator is used. The source behaviour is specified by the distribution of the idle time of the represented source and the holding time of a connection (Figure 2). If a setup request has to be refused due to blocking in the node the source may be blocked for some time. After this period the call attempt is repeated after the repeat time. The distributions of the blocking and the repeat time can also be defined.

This source model is in contrast to most analytical approaches for evaluating node architectures described in literature, which use Poisson traffic with negative-exponentially distributed interarrival times and holding times ("pure chance traffic") [2]. Usually however, this simplified model holds only for a quasi infinite number of sources and is not a very good approximation for a smaller number of sources as expected in photonic networks. Especially, we expect the repeat behaviour of a wavelength channel to be different from "classical" sources due to the high traffic streams transported on a single wavelength. Dividing the time between a refused call attempt and the repetition of this call attempt into two phases has an additional advantage (besides the possibility of exactly modelling a real source). If the distribution types for holding and blocking time are chosen identically, the total offered load does not depend on the blocking values and therefore a comparison with analytical results can be done more easy.

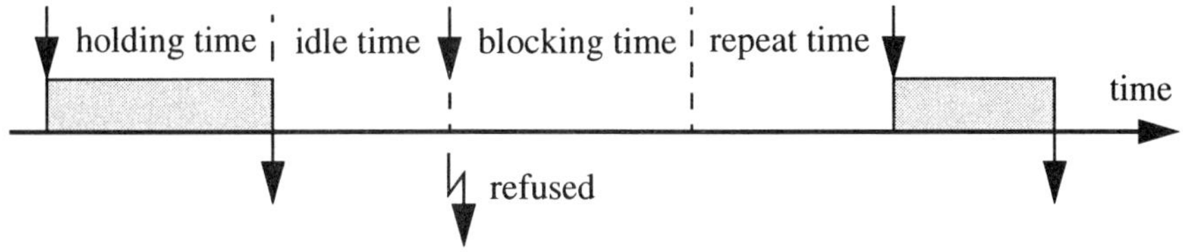

Figure 2 Source model used for node architecture evaluation

Apart from the local traffic input links, our model considers additional input links at the node which represent the traffic coming from other nodes. In this way it is also possible to build up an arbitrary concatenation of multiple nodes. Moreover, the architecture of the cross-connect defines its functionality and therefore determines the modelling of the "black box" shown in Figure 1. When modelling this part, aspects like the availability of wavelength conversion or the functionality of the applied switching stage inside the node have to be considered.

For larger networks with many channels per link however, this approach leads to a high number of generators in the simulation model. Therefore, for investigating network aspects and especially the influence of different routing strategies on network performance a different approach is applied. This approach uses one traffic generator to characterise the traffic between one pair of source and sink with appropriate distribution functions for the interarrival times of call requests and the call holding time, respectively (Figure 3). For the parameterization of the generators it is also possible to use measurement data.

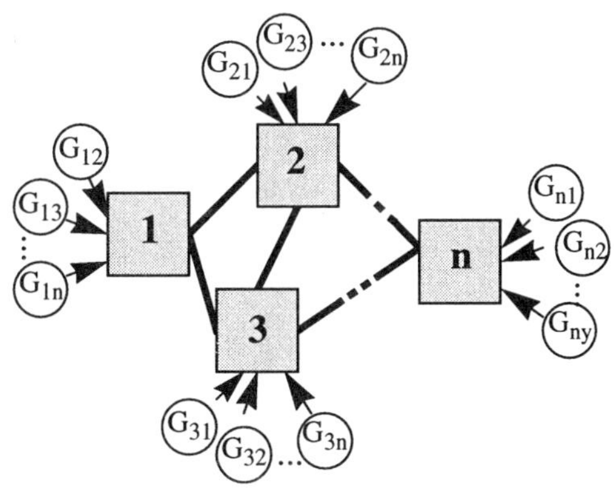

Figure 3 Modelling approach for investigating routing strategies

For the modelling of routing methods we also use two different modelling approaches:

- For *hop-by-hop routing* methods, each node has a separate routing table which contains the alternatives for the next hop. The decisions for the next path section are made locally in every node depending on the destination node. If a node cannot find an alternative, the call is either blocked or control is given back to the predecessor node (*crankback*).

- For *source routing* methods, each source contains a table with the full routing information for the path to any destination node. All decisions concerning the route choice are made only in the source node. This approach also covers all methods based on decisions made by a central network management (for example TMN).

The described modelling approaches form the base for our discrete event simulation tools which are realised in an object-oriented way in C++. The tools allow a flexible definition of traffic characteristics. The topology of the investigated network can be chosen arbitrarily. This includes the placement and interconnection of nodes as well as the dimensioning of link capacities in terms of numbers of fibres. The routing strategy is described with an input parameter. If available, the routing alternatives can be defined freely. For the node architecture, several possibilities are already implemented, including the two extreme cases nodes *without* converters and nodes *fully* equipped with converters. Many other concepts can be thought of and are left for further studies.

The simulation tools provide a variety of results. Examples are link load or blocking probabilities for every link, every node, every source-destination node pair, or the whole network. The results are available as mean values with confidence intervals or - where appropriate - as distribution histograms.

4 RESULTS AND CASE STUDIES

4.1 Evaluation of node architectures under dynamic traffic load

The investigation of more complex node architectures usually requires simulative investigation methods. In the following, simulation results are presented for a CC with 8 input and 8 output links (with one or two fibres) including links for local traffic. Each output link is chosen with the same probability, i.e., no specific routing strategy is considered. We assume a share-per-node architecture where the available converters can be shared by all connections. The main goals are to evaluate the influence of the number of converters, of available WDM channels per fibre, and of fibres per link on the node performance, respectively. The offered load per wavelength and link is kept constant (0.5 Erlang), i.e., the total load is 32 Erlang for 8 wavelengths per fibre (8 WL), or 64 Erlang for 16 wavelengths per fibre in the case of one fibre per link (in case of two fibres per link these values are already achieved with 4 or 16 wavelengths, respectively).

First, the results confirm the effect that already some converters improve the node performance significantly whereas from a certain point on further converters cannot decrease blocking any more (the lines are horizontal in Figure 4). Moreover, the well-known effect of "economy of scales" can also be seen here: with a higher total number of wavelength channels lower blocking values are achievable. However, the total number of converters required to achieve the minimum blocking is higher

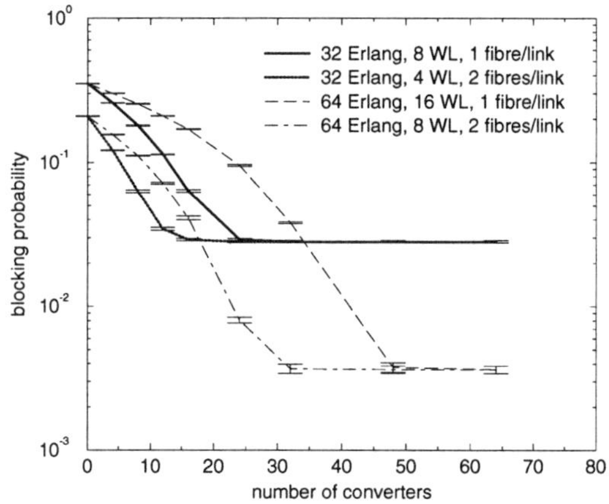

Figure 4 Blocking probability for a node with 8 input and 8 output links

than for smaller systems. Another possibility to improve the performance is to use multiple fibres - each with fewer wavelengths - for each link. In that case, fewer converters are required to decrease blocking significantly.

Other investigations showed that the quantitative performance improvement also strongly depends on the offered traffic load. With higher load, only small improvements can be achieved because then blocking due to overload is the main reason. Looking at the absolute blocking values, we see that for the assumed traffic dynamics blocking is relatively high. This leads to the conclusion that for optical cross-connecting we need either relatively high over-dimensioning, or the traffic dynamics have to be very small, or we have to reach larger bundles, i.e. bigger systems where blocking is improved due to economy of scales.

4.2 Evaluation of routing strategies

We concentrate in the following on the WP and the VWP case. The results are derived for an example network with nine nodes which could represent a national German network. The traffic demands (called *static traffic demand* in the following) were derived from population size and node distances applying simple rules. The network is dimensioned according to these static traffic demands using the shortest paths and then investigated under dynamic traffic conditions. Although our tools could also handle uni-directional calls, we do not focus on this and use only symmetrical bi-directional calls. Moreover, negative-exponential distributions are used for interarrival and call holding times.

In Figure 5, several routing methods are compared for the WP case. The first method is source routing based on the paths determined during network dimensioning (here: shortest paths) without alternatives. The other methods are hop-by-hop methods: with *OOC* (Originating-Office Control) only the originating node is able to choose an alternative path whereas all other nodes have to choose the first routing possibility towards the end node. In our case this means that each node checks all available wavelength channels for the primary direction. With *SOC* (Sequential-Office Control), each node is able to choose between a number of alternatives (i.e., all wavelength channels in multiple directions), but if a node is not able to find any way for the next path section the call is blocked. If *SOC with crankback* is used, then in case of blocking routing control may be returned to the predecessor which can now try to find other alternatives. Therefore, this is the most powerful method of the considered ones whereas fixed source routing is the most restricted one.

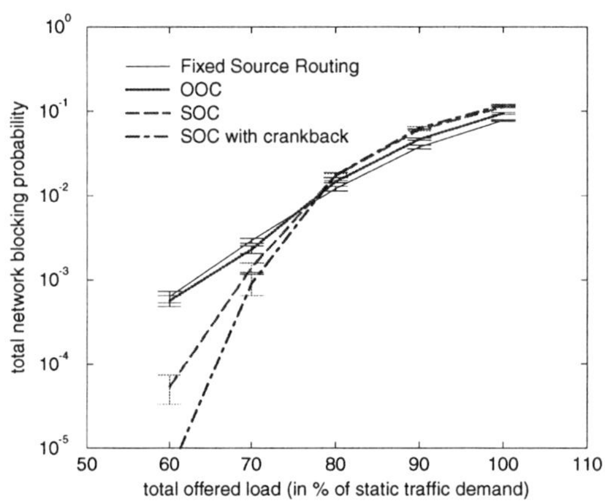

Figure 5 Comparison of routing methods in a WP network (no converters)

Figure 5 shows that the behaviour of the methods depends on the load. While for low load the methods which are able to choose between many alternatives lead to lower blocking probabilities, for high load the fixed source routing method which allows no alternatives is best. The reason is that alternatives usually lead to longer paths and therefore require more network resources. So alternate routing only leads to better overall performance if blocking occurs seldom. This effect is increased in the WP case considered here where a continuously free wavelength from source to destination has to be found.

Figure 6 compares two different routing strategies for WP and VWP respectively. Again, we see the differences between low and high load. Furthermore, the figure shows that wavelength converters are especially useful with sophisticated routing strategies: the improvement for *SOC with crankback* compared to fixed routing is much better in the VWP case than in the WP case.

Another result which can be derived from these figures is that a network planned for a certain static traffic capacity is not able to handle this traffic under dynamic changes, too. If dynamic traffic is present, the mean value of offered load has to be significantly lower than the static traffic demand to achieve acceptable blocking probabilities.

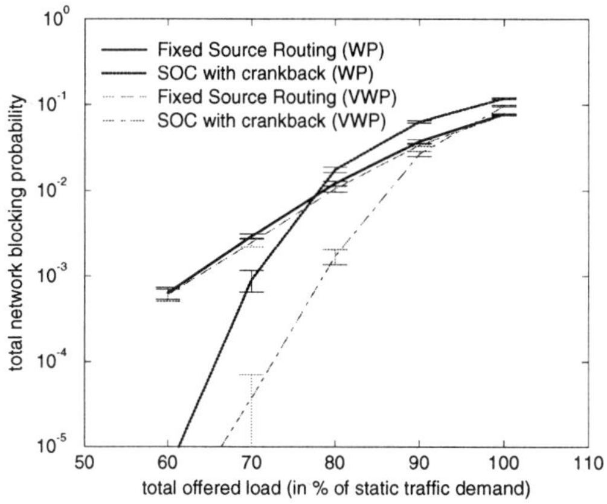

Figure 6 Comparison of different routing methods for WP and VWP case

5 CONCLUSIONS AND OUTLOOK

In this paper, we presented two modelling approaches for the simulative investigation of network performance under dynamic traffic conditions in an optical transport network. The developed tools based on these models allow performance investigations for arbitrary node and network structures for different traffic types. First results were derived for a single node showing the influence of the number of converters, the number of fibres, as well as the number of WDM channels per fibre on the node performance. Moreover, several routing strategies and their influence on network performance were evaluated.

Future work will focus on applying enhanced traffic models and on modelling the node architecture in more detail. Moreover, multi-layer aspects should be taken into account since the traffic streams in the optical layer result from aggregating various higher layer traffic streams. Finally, other routing methods will be evaluated considering more the specific features of an optical layer.

ACKNOWLEDGEMENT

This work has been supported by the German BMBF (Bundesministerium für Bildung, Wissenschaft, Forschung und Technologie) under contract 01BP504/2. The authors thank D. Porbadnigk and M. Koch for implementing the simulation tools.

REFERENCES

[1] A. Girard: *Routing and Dimensioning in Circuit-Switched Networks*; Addison-Wesley, 1990

[2] J. Späth et al., Architecture and Performance Evaluation of Future Photonic Networks, *Proceedings of the 12th ICCC*, Seoul, 1995, Vol. 2, pp. 661 - 666

PART TWO

Wavelength conversion and assignment

Devices for all-optical wavelength conversion and spectral inversion

Antonio Mecozzi

Fondazione Ugo Bordoni, via B. Castiglione 59, I-00142 Roma, Italy

Phone: +39 (6) 5480--2232 Fax: +39 (6) 5480--4402

Abstract

Recent progress towards efficient frequency converters and spectral inverters based on semiconductor optical amplifiers is reviewed. Criteria for the optimization of the amplifier structure for the particular use are given. Special attention is paid to techniques based on four-wave mixing used for both wavelength conversion and dispersion compensation.

Summary

The last few years have witnessed a tremendous increase in the capacity of point-to-point optical communication systems, both in laboratory experiments and in deployed systems. It is not clear at the moment how much of the bandwidth potentially available from the point-to-point technology will be immediately exploited at the network level. One of the main limitations to the full use of the poin-to-point bandwidth is the speed of the devices used for the processing at the network nodes. Although other options are currently investigated, it is generally acknowledged that a higher capacity will be achieved by the use of devices that process the signal at the optical level, with no need of electronic regeneration. Since the main limit to a high network capacity is the accumulation of noise and signal distorsion induced by the nonlinearity, it is easy to anticipate that the ultimate network capacity will be reached by the use of optical R^3 regenerators, capable of reshape, retime and reamplification of the signal. Many applications, however, require a linear transfer function. For linear transfer function, we mean that the amplitude of the output signal is proportional to the amplitude of the input signal, and the its phase is either equal or opposite to the phase of the input. Among the applications requiring linearity of the transfer function of the device, we mention the simultaneous processing of many wavelength division multiplexed channels or the processing of analog signals. Regeneration of the signal is obviously impossible by linear devices, as reshaping necessarily requires a nonlinear tranfer function.

In this presentation, we will review the recent progress towards the realization of devices for all optical signal processing. Two functionalities will be mainly discussed. One is wavelength conversion, and we will focus on device based on four-wave mixing and cross-phase and cross-gain modulation in semiconductor optical amplifiers. The other is spectral inversion based on four-wave mixing in semiconductor optical amplifiers for dispersion compensation by mid-span spectral inversion. Devices based on four-wave mixing have linear transfer function and are fully transparent to the modulation format, those based on cross-gain and cross-phase modulation have nonlinear transfer function and are not transparent. On the other hand, some of the devices based on cross-phase modulation in interferometric configurations produce a partial reshaping of the signal, although not its retiming.

It is generally acknowledged that semiconductor optical amplifiers are promising candidates as efficient and compact devices for four-wave mixing applications. The converter performance is characterized by the efficiency and by the noise added by the conversion process. We define the efficiency as the ratio between the power of the converted output $P_2(L)$ and the power of the input signal $P_1(0)$. The efficiency of FWM in semiconductor optical amplifiers has already been studied in details [1,2] In Ref. [2] the simple formula for the efficiency was given

$$\eta = \frac{P_2(L)}{P_1(0)} = f|R(\Omega)|^2 \left(\ln \frac{G_0}{G} \right)^2$$

where $f = 1/[1+P_1(0)/P_0(0)]^2$, $P_0(0)$ is the pump power, $R(\Omega)$ is the material response function, which depends on the parameters characterizing the semiconductor nonlinearity and on the frequency detuning between pump and signal $\Omega/(2\pi)$ and G_0 and G are the unsaturated and the saturated gain of the amplifier. By this equation, it is possible to show the existence of a pump power corresponding to a maximum of the efficiency. Furthermore, it is possible to optimize the device parameters to get high conversion efficiency. In Ref.[3], by using a device 1 mm long, a conversion efficiency larger than 0 dB was reported for a conversion interval (the detuning between the input and converted signal) as large as 2 THz for frequency upconversion, 1 THz for frequency downconversion. With a 1.5 mm and 2.0 mm devices, the efficiency and the noise characteristic of the device are even better [4,5]. It should be noted that the linear gain of semiconductor optical amplifiers does not indefinitely grow with distance as it might be expected, it shows instead a tendency to saturation. This effect is caused by the increasing gain saturation caused by the broadband amplified spontaneous emission. For this reason, amplifiers longer than few hundreds of microns have never been fabricated for linear applications. A semiconductor optical amplifier used for four-wave mixing is however always operating under strong gain saturation from the optical pump, and it is therefore expected that the efficiency grows with the device length, despite the fact that the linear characteristics degrade. Work at Fondazione Ugo Bordoni lead to the prediction that long amplifiers have high performance for nonlinear applications. As an outcome of a collaboration with FUB, the small Swiss manifacturer OptoSpeed has been the first to produce and to make commercially available long amplifiers for this particular use. To date, also Alcatel produces amplifiers longer than 1 mm specifically designed for nonlinear operation. The world record performance in terms of both efficiency and signal-to noise ratio of a four-wave mixing converter is presently detained by a 2 mm long multi-quantum-well device fabricated by British Telecom [6].

Another technique for frequency conversion is based on cross-gain and cross-phase modulation. Semiconductor optical amplifiers in interferometric configurations have shown excellent performance for frequency conversion. Work at FUB in collaboration with British Telecom has shown that, surprisingly, the waveguide internal loss is beneficial to increase the extinction ratio of the frequency converted signal [7,8]. Recent work at Alcatel has shown that an optimization of the amplifier performance as frequency converter is indeed possible by a careful engineering of the internal loss of the amplifier [9].

The most promising application of four-wave mixing in semiconductor optical amplifiers is dispersion compensation, obtained inverting the spectrum of an optical signal midway between the transmitted and the receiver. One of the problem of mid-span spectral inversion is the waste of bandwidth caused by the fact that the converted signal is centered at a different frequency than the input. A method to obtain spectral inversion without frequency shift based on the use of two orthogonal pumps in a semiconductor optical amplifier has been recently proposed at FUB [10,11]. Independently, K. Inoue at Nippon Telephon and Telegraph came out with essentially the same scheme using four-wave mixing in optical fibers [12].

References

[1] J. Zhou, N. Park, J. W. Dawson, K. J. Vahala, and M. A. Newkirk, "Efficiency of broadband four-wave mixing wavelength conversion using semiconductor traveling-wave amplifiers," IEEE Photon. Technol. Lett. **6**, 50–52 (1994).

[2] A. Mecozzi, "Analytical theory of four-wave mixing in semiconductor amplifiers," Opt. Lett. **19**, 892–894 (1994).

[3] A. D'Ottavi, F. Martelli, P. Spano, A. Mecozzi, S. Scotti, R. Dall'Ara, J. Eckner, and G. Guekos, "Very high efficiency four-wave mixing in a single semiconductor traveling-wave amplifier," Appl. Phys. Lett. **68**, 2186–2188 (1996).

[4] F. Girardin, J. Eckner, G. Guekos, R. Dall'Ara, A. Mecozzi, A. D'Ottavi, F. Martelli, S. Scotti, and P. Spano, "Low-noise and very high-efficiency four-wave mixing in 1.5-mm-long semiconductor optical amplifiers," IEEE Photon. Technol. Lett. **9**, 746–748 (1997).

[5] A. D'Ottavi, F. Girardin, L. Graziani, F. Martelli, P. Spano, A. Mecozzi, Scotti, R. Dall'Ara, J. Eckner, and G. Guekos, "Four wave mixing in semiconductor optical amplifiers: a practical tool for wavelength conversion," IEEE J. Sel. Top. Quantum Electron. **3**, 522–528 (1997).

[6] D. D Marcenac, D. Nesset, A. E. Kelly, M. Briely, A. D. Ellis, D. G. Moodie, and C. W. Ford, “40 Gbit/s transmission over 406 km of NDSF using mid-span spectral inversion in a 2 mm long semiconductor optical amplifier,” Electron. Lett. **33**, 879–881 (1997).

[7] A. Mecozzi, “Small signal theory of wavelength converters based on cross-gain modulation in semiconductor optical amplifiers,” IEEE Photon. Technol. Lett. **8**, 1471–1473 (1996). A. Mecozzi and J. Mørk, “Saturation effects in nondegenerate four-wave mixing between short optical pulses in semiconductor optical amplifiers,” J. Sel. Topics Quantum Electron. **3**, 1190–1207 (1997).

[8] D. Marcenac and A. Mecozzi, “Switches and frequency converters based on cross-gain modulation in semiconductor optical amplifiers,” IEEE Photon. Technol. Lett. **9**, 749–751 (1997).

[9] T. Ducellier, “Integrated GAG-SOA (Gain-Attenuation-Gain) for ultra-high speed wavelength conversion,” SOA based components for optical networks workshop, Praha, October 27-28, 1997.

[10] A. Mecozzi, Italian patent No. RM 96 A 000786, November 18, 1996.

[11] A. Mecozzi, G. Contestabile, F. Martelli, L. Graziani, A. D'Ottavi, P. Spano, R. Dall'Ara, and J. Eckner, F. Girardin, and G. Guekos, “Optical spectral inversion without frequency shift by four-wave mixing using two pumps with orthogonal polarizations,” IEEE Photon. Technol. Lett. **10**, 335–337 (1998).

[12] K. Inoue, “Spectral inversion with no wavelength shift based on four-wave mixing with orthogonal pump beams,” Opt. Lett. **22**, 1772–1774, (1997).

Where to locate wavelength conversion in WDM networks ?

M.Garnot, P. A. Perrier

Alcatel Alsthom Corporate Research Center
c/o Alcatel Alsthom Recherche
Route de Nozay - 91460 Marcoussis (France)
tel : (33) 1 69 63 13 60
fax : (33) 1 69 63 18 65
e-mail : Mathieu.Garnot@aar.alcatel-alsthom.fr

Abstract

The massive deployment of Wavelength-Division Multiplexing (WDM) in existing transport networks worldwide will facilitate the definition of a photonic transport layer. Different routing strategies of optical paths in such networks have been investigated [1] ; resource allocation has also been proposed using two different schemes based on wavelength routing and wavelength switching, respectively. The design of a WDM transport network not only implies identifying algorithms for routing and resource allocation but also modeling of different routing functions at the node and network levels. Such a function is wavelength conversion.

This paper addresses the use of wavelength conversion in WDM networks.

Introduction : A transport network overview

The introduction of WDM [2,3] in telecommunication networks is today a worldwide concern. WDM is perceived as a simple, modular and cost-effective technique to enhance the capacity of existing networks through a more efficient use of the available bandwidth of optical fibers.

In addition to the mere capacity increase it offers, the extensive deployment of WDM point-to-point transmission systems opens up the possibility to efficiently perform networking functions directly in the optical domain. With each of the transported signals assigned a distinct wavelength (wavelength channel), carried from node to node on the same fiber, it is possible using space-division and wavelength-division switching techniques to route each channel individually without any synchronization or opto-electronic conversion. New optical routing elements, such as Optical Add-Drop Multiplexers (OADMs) and Optical Crossconnects (OXCs), can thus be designed to only extract channels that have reached their destination, and let the others, possibly multiplexed with locally added channels, continue transparently to the following nodes. Many experimental implementations of OXCs and OADMs have already been demosntrated [4] which can be classified into two major classes : Wavelength Switching or Translating (WT) and Wavelength Routing (WR), depending whether wavelength converters are used or not.

General benefits of wavelength conversion

The difference between WR and WT nodes is depicted in Figure 1. In a WR network an optical path is assigned a single wavelength from source to destination. Given that the set of available wavelengths is, in practice, limited (e.g., 32), these wavelengths will have to be reused throughout the network to establish all the required connections provided that two optical paths at the same wavelength do not share a common fiber. In this scheme, wavelength contention can be alleviated only by allocating the optical paths with a global view of all the connections.

In a WT network, wavelength is considered a physical transport resource with a local significance which can be reused from link to link between source and destination. An optical path is therefore assigned – not a single wavelength,– but a concatenation of several, possibly different, wavelengths. Here, wavelength contention can be solved simply by performing wavelength at the nodes.

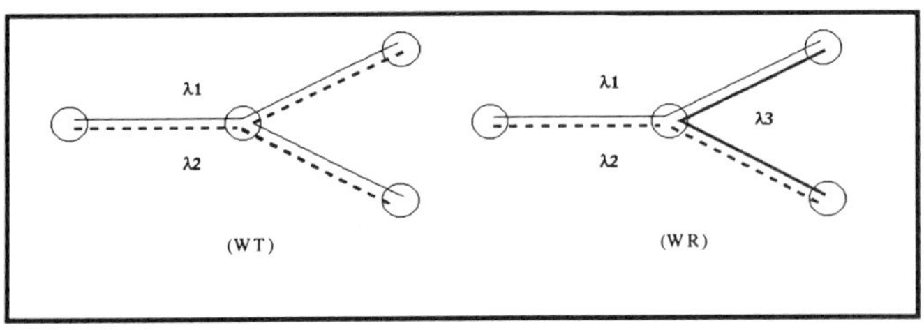

Figure 1:Wavelength Routing (WR) vs
Wavelength Translating (WT) networking schemes

The extent to which wavelength translation is required has given rise to heated debates. Intuitively, however, wavelength conversion can save resources (fewer fibers and smaller nodes) and simplify the management of optical networks [1,5]. The gain to be expected will depend on the network's characteristics (topology, number of nodes, diameter, etc) and the supported traffic (of the optical layer). The results, reported below, on a Pan-American Optical Network [6] show the benefits of wavelength translation [1] in the case of a real network. Though general analytical expressions have been given for regular topologies [7], such expressions for existing networks remain to be determined. Nevertheless, some trends can be derived and explained. For example, it has been found that wavelength conversion is more useful in large-scale or low-connectivity networks when the number of wavelengths is large or when traffic is changing dynamically.

From a physical point of view, it was shown that some all-optical wavelength converters regenerate the Optical Signal-Noise-Ratio. This specific feature can be very attractive in order to optimize the WDM network design. However, the number of all-optical wavelength converters that can be cascaded will be limited by jitter accumulation introduced during the conversion process. This actual drawback concerns only the all-optical implementation of the wavelength conversion ; other implementations (i.e., opto-electronic) can be envisioned that do not suffer from this drawback.

Modeling of wavelength conversion

According to the traffic, the network topology, and the protection schemes [8], the dimensioning process establishes the end-to-end optical paths and the associated resources. The (optical) traffic matrix is derived from the transport traffic matrix by grooming and multiplexing. The optical network topology comprises all the optical network elements which are involved in the transport and routing functions of the WDM layer. We define this topology as a graph in which the vertices represent the optical routing nodes and the axes the fiber links.

The problem is then to assign the appropriate resources to the given paths. It can be shown that resource assignment is an NP-complete (Non-deterministic Polynomial [9,10]) problem when wavelength converters are used everywhere in the network. Actually, the severe constraint of wavelength continuity (WR) when wavelength conversion is not used turns the wavelength assignment problem into a graph colorability problem [11], which is NP-complete. The problem is not NP-Complete only when wavelength conversion is used at every node in the network. Therefore, introducing wavelength conversion only in a few routing nodes does not change the nature of the allocation problem. As a matter of fact, and paradoxically, the addition of wavelength conversion can result in an increase of the number of required resources (fiber, node size, etc).

Unfortunately, for cost reason and optimal benefits, the use of wavelength conversion will be probably restricted to a few routing nodes. Also, we apply the classical wavelength allocation algorithm [11] in order to assign wavelengths to sections of the optical paths (sub-paths). A sub-path is defined as the part of an optical path which does not cross wavelength translating nodes. The following figure shows an end-to-end optical path defined by the traffic matrix and the different sub-paths determined by the addition of wavelength conversion.

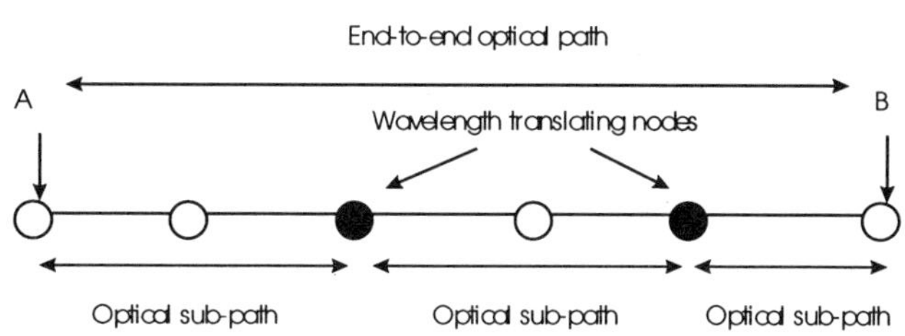

Figure 2 : Paths and sub-paths

In the wavelength allocation algorithm, each sub-path is treated individually ; the wavelength assigned to that sub-path can be different from the wavelengths assigned to the other sub-paths even though all sub-paths belong to the same end-to-end path.

How to select the nodes with wavelength conversion

The locations at which wavelength conversion will take place is a key point to investigate. Actually, the criteria to select these locations are based on wavelength blocking and on the number and length of the generated sub-paths.

Wavelength blocking can be estimated by analyzing different traffic configurations. This analysis leads to different mathematical models which are not always practical to rapidly obtain dimensioning results. Therefore, we have defined simple criteria to choose the best suited location of wavelength converting nodes, resulting in a better understanding of the global behavior of the WDM optical layer.. These criteria, applied to each routing nodes, are :

- criterion A : number of optical channels in transit
- criterion B : (number of optical channels in transit)*(node connectivity)
- criterion C : output fiber utilization ratio

The number but more importantly the length, determined by the number of used links, of the generated sub-paths imposes the introduction of wavelength conversion in the longer (sub-) paths. It follows that criterion A can be modified into criterion A' : number of optical channels in transit corresponding to an optical (sub-) path longer than a minimum from end-to-end.

Application and results

The test network (Figure 3) is the Pan-American Network [6] which interconnects the main cities (maximum number of nodes=78) in the United-States. This example is interesting since it proposes a large scale network with different traffic configurations. Actually, the traffic is mainly transversal (from East to West and West to East), with high density around the large metropolis (Chicago , New-York, Dallas, ...).

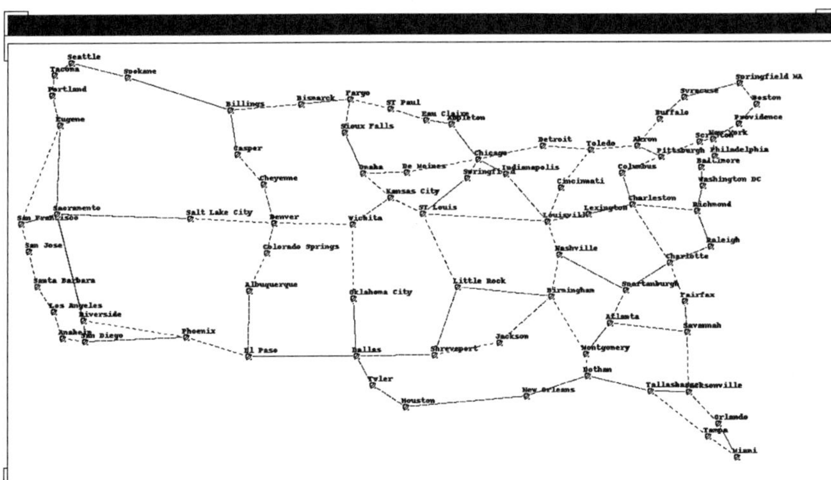

Figure 3 : Pan-American network

Using optical cross-connects without wavelength conversion, the dimensioning results lead to about 88 000 miles of fiber (maximum of 8 wavelengths per fiber ; maximum bit rate of 2.5 Gbit/s on each wavelength) for a traffic matrix containing 466 connections.

The results when the different criteria are applied are presented in figure 4, where the total fiber length is given as a function of the number nodes fully equipped with wavelength converters.

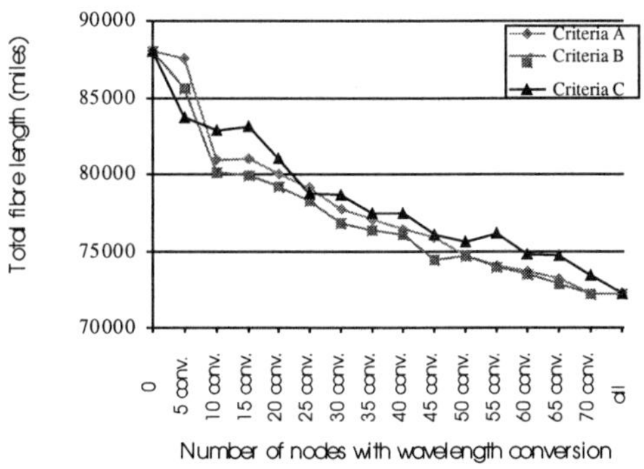

Figure 4 : Total fiber length as a function of the number nodes with wavelength conversion

The figure shows first that the application of the different criteria leads to a significant reduction (about 15%) in the total fiber length required, and second that some criteria are more efficient (less number of nodes with wavelength conversion for a given reduction of the fiber length). For example by introducing 15 nodes with wavelength conversion, there is a difference of about 3 000 miles (3 to 4%) between criteria B and C.

It also shows the possible increase of the required fiber length when more nodes with wavelength conversion are introduced (from 10 to 15 nodes with criterion C). This behavior illustrates the NP-completeness of the wavelength allocation problem when wavelength conversion is not used everywhere in the network.

Conclusion

This paper briefly reviews the interest of the WDM technology introduction in transport networks. Our study focuses on the use of wavelength conversion, one of the most attractive feature of the future WDM transport layer. Different criteria have been proposed to select the best locations to introduce this function. It was noticed that the wavelength allocation problem is always NP-complete if only a few routing nodes use wavelength conversion. For this reason, the criteria are based on wavelength blocking and wavelength allocation algorithm with the taget to minimize the number of long sub-paths. We found that the network connectivity is one of the most important parameter (criterion B), but no theoretical formalism can be proposed to this date.

References

[1] M. Garnot et al., "Routing strategies for optical paths in the WDM networks", ICC'97, Montreal, June 8-12,, pp. 422-426.

[2] C. A. Brackett, "Dense wavelength division multiplexing networks: principles and applications", IEEE J.on Sel. Areas in Comm., vol. 8, no. 6, pp. 948-964, Aug. 1990.

[3] P. Demeester et al., "Photonic transport networks: why, how and when ?", NOC'96, vol. 1, Heidelberger, June, pp. 158-165.

[4] A. Jourdan et al., " Design and implementation of a fully reconfigurable all-optical crossconnect for high capacity multiwavelength transport network", IEEE J. of Lightwave Technology, vol. 14, no. 6, pp. 1198-1206, June 1996.

[5] N. Wauters et al., "Reduction of wavelength blocking through partitioning with wavelength converters", OFC'97, Dallas, Feb., pp. 122-123.

[6] M.Garnot et al., " Dimensioning and design of the WDM optical layer in transport network" , SPIE conference on All-Optical Communication : Architecture, Control, and Network Issues III, November 2-3, 1997, pp. 244-252.

[7] R. A. Barry, P. Humblet, "Models of blocking probability in all optical networks with and without wavelength changers", IEEE INFOCOM'95, Boston,.

[8] L. Nederlof et al., "A comparison of survivability methods for the optical layer", NOC'97, June 17-20, Antwerp, Belgium, pp. 103-110.

[9] M. R. Garey and D. S. Johnson, Computers and intractablility: a guide to the theory of the NP-Completeness, Freeman Publishers, 1979.

[10] B.M.E. Moret and H.D. Shapiro, Algorithms from P to NP, The Benjamin / Cummings Publishing Company, 1991.

[11] I. Chlamtac et al., "Lighpath Communication: An Approach to High Bandwidth Optical WAN's", IEEE Transactions on Communications, vol. 40, no. 7, July 1992, pp. 1171-1182.

Wavelength Assignment in WDM Networks – A Comparison of Approaches

B. Mikac

R. Inkret

*F. Tillerot**

Faculty of Electrical Engineering and Computing

Unska 3, HR-10000 Zagreb, Croatia

Phone: +385 1 6129 752, +385 1 6129 658

Fax: +385 1 6129 832

Email: branko.mikac@fer.hr, robert.inkret@tel.fer.hr

**France Telecom,*

2 Avenue Pierre Marzin

F-22307 Lannion Cedex, France

Phone: +33 296 05 10 73, Fax: +33 296 05 10 08

Email: francois.tillerot@cnet.francetelecom.fr

Abstract

Two heuristic search techniques, genetic algorithm and simulated annealing, together with a simple heuristic procedure, based on the graph coloring, are applied in searching for optimal solution of the wavelength assignment problem in all-optical transmission networks with wavelength division multiplex (*WDM*). The assignment results are compared.

Keywords

Genetic Algorithm, Simulated Annealing, Graph Coloring, Wavelength Division Multiplex, Wavelength Assignment

1 INTRODUCTION

The wavelength assignment as optimization problem appears in all-optical transmission networks with wavelength division multiplex (*WDM*). An optimization goal could be defined as: how to assign wavelengths, fibers and optical links in order to minimize total fiber length (*TFL*) in the network, fulfilling traffic and reliability requirements.

The paper describes an application of genetic algorithm (*GA*), simulated annealing (*SA*), and a heuristic procedure based on graph coloring (*GC*), in order to compare methods according to optimization results.

2 WAVELENGTH ASSIGNMENT IN WDM NETWORKS

In *WDM* a number of channels (N_λ) are transmitted on a single fiber by using different wavelengths, referred to as wavelength channels (*WCH*). A single *WCH* carries the traffic of standard bit rates, connecting a pair of nodes by using an optical path. Each *WCH* is defined by a sequence of triples {link [fiber (wavelength)]}. In real transmission systems, the number of *WCH*s multiplexed onto a single optical fiber is constrained. In order to fulfil traffic demands, the number of available optical fibers on an optical link is assumed unconstrained. The wavelength assignment should be done in such a way to avoid a space (fiber), wavelength, or time conflict between pairs of triples.

From the availability reasons, the path protection scenario is assumed. Each *WCH* could be carried by a primary path (P_x), the first shortest path used in faulty free condition, or by its spare path (S_x), the second shortest independent path used in the faulty condition of any element on the primary path.

In this analysis, the following assumptions are made: A *WCH* uses a single wavelength along all fibers on its route. Primary and spare *WCH*s could be assigned to different wavelengths.

The correctness of wavelength assignment could be checked according to the following rule:

- Two *WCH*s could be assigned to the same wavelength if they are mutually independent (*I*), or disjoint (*D*). Two *WCH*s are defined as mutually independent if they are carried by the paths that have no fiber in common. Two *WCH*s are disjoint as events, if both can not occur at the same time under assumption of a single element failure.

Under proposed protection scenario with single failure tolerance assumptions, the previous rule has the following consequences:

- Two primary paths are never disjoint.
- A primary path and its spare path are always independent and disjoint.
- A P_x and a S_y, are disjoint and independent, if $P_x \subseteq P_y$.
- Two spare paths are disjoint if their primary paths are independent.

An example on Figure 1 shows the relationship between *WCH*s in sense of independence and disjointedness. Two pairs of *WCH*s connect node pairs *A*,*C* and *B*,*C* respectively, by using primary and spare paths. In this case the same wavelength could be assigned to all *WCH*s, because P_x and P_y are independent, and S_x and S_y are disjoint under single failure assumption.

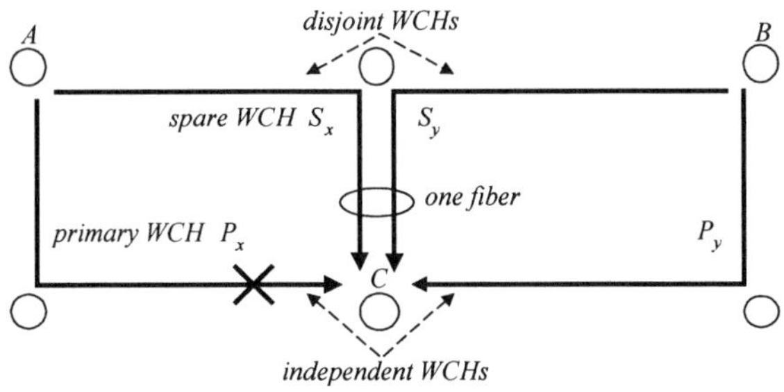

Figure 1 An example of independent and disjoint wavelength channels

In order to minimize the total fiber length, the minimization is done in two steps:

(1) Wavelengths are assigned to all *WCHs*.

(2) The fiber-wavelength assignment for each optical link is done on the following basis: a *WCH* w_k with the wavelength λ_i is assigned to the first fiber

  (a) on which λ_i is not occupied at all, or

  (b) if the appearance of w_k is disjoint to the appearance of all already assigned *WCHs* w_j, $j \neq k$, which have wavelength λ_i.

In optimization procedure the following formulas were used:

$$f = \sum_{k=1}^{E} \left[F_k l_k + \alpha \frac{1}{UF_k} \right] \tag{1}$$

where f is the goal function to be minimized, E is the number of links in the network, F_k is the number of optical fibers on the optical link k, l_k is the length of the link k, w_k is the number of used wavelength channels on the link k, and α is real constant which determines the contribution size of utilization factor UF_k to goal function, for each link k. The utilization factor UF for entire network, which shows the fraction of total network capacity occupied by assigned *WCHs*, can be calculated by following expression:

$$UF = \frac{1}{E} \sum_{k=1}^{E} UF_k = \frac{1}{E} \sum_{k=1}^{E} \frac{w_k}{F_k l_k} \tag{2}$$

3 APPLICATION OF GENETIC ALGORITHM AND SIMULATED ANNEALING

Both *GA* [Goldberg] and *SA* [Aarts] belong to the class of general heuristic algorithms rooted in natural processes.

A modified simple *GA* [Goldberg] was used with the following parameters: one-point crossover, tournament selection technique, population size=100, crossover probability=0.6, mutation probability=0.01, and elitism.

In *k*th iteration of used *SA* algorithm, the temperature, as a control parameter, is decreased according to the following rule $c_k = \alpha c_{k-1}$. The number of examined solutions increases according to the following formula $L_k = L_{k-1}/\beta^{0.1}$. The actual number of transitions executed in kth iteration is $L_k^* = \min_k\{L_k, L_{max}\}$, where L_{max} is the maximum number of accepted transitions.

The values of constants defining cooling schedule are as follows: α=0.95, β=1.01, c_0= 300, L_0=200, and L_{max}=100.

In *GA* and *SA*, the solutions are encoded as arrays of integers, where each integer specifies a wavelength of corresponding *WCH* using $\log_2 N_\lambda$ bits per each integer. Figure 2 shows an example of coding 5 *WCH*s with maximum number of N_λ=4 wavelengths.

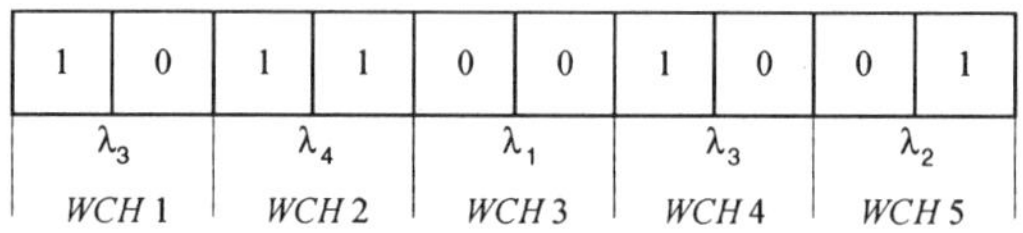

Figure 2 Solution coding in *GA* and *SA*

If the total number of *WCHs* in a network is N_{WCH} then the number of all potential solutions is

$$N_{TOT} = 2^{N_{WCH} \times \beta \times \log_2(N_\lambda)}. \qquad (3)$$

The factor β depends if the primary and its spare path have the same wavelength (β=1), or the primary and its spare path can have different wavelengths (β=2).

4 WAVELENGTH ASSIGNMENT AS COLORING PROBLEM

In order to decrease very high complexity of wavelength assignment problem independence/disjointedness tests (*I/D* tests) are used, as defined in Chapter 2. The results of *I/D* tests can be represented by a *wavelength relationship graph WRG*. Each node in the *WRG* represents a *WCH* to which a wavelength should be assigned. A branch between a pair of nodes represents the fact that these nodes can not be assigned to the same wavelength. The problem of optimal wavelength assignment - how to assign wavelengths to *WCH* using minimal number of wavelengths - could be transformed to the graph coloring problem, where one should color the *WRG*, using minimal number of colors. Every regular graph coloring gives a regular solution of wavelength assignment problem, as compared to *GA* or *SA*, where a very big fraction of evaluated solutions are irregular. In searching for optimal graph coloring an *NP*-complete problem should be solved. In searching for a quasi-optimal graph coloring we use simple heuristic algorithm [Welsh, Powell].

(1) In iteration k of algorithm, an uncolored node with highest node degree is colored by the color (wavelength) k. In the same iteration, additionally, all uncolored nodes, not neighboring to k-colored nodes, are colored, step-by-step, by the color k as well.

Figure 3 shows an example of wavelength assignment in a *WDM* network with 4 nodes and 5 links. Each link has 2 fibers for both directions. Each node pair is connecting by 2 *WCHs*, primary *P*, and spare *S* (see Table 1). The results of wavelength assignment by using wavelength relationship graph and simple coloring algorithm are presented in Table 1, as well.

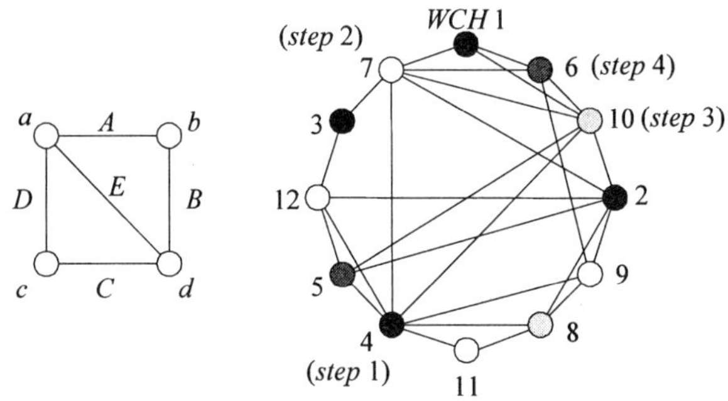

Figure 3 An example network and its wavelength relationship graph (*WRG*) for the routing arrangement as mentioned in Table 1

Table 1 Wavelength assignment results for example network

Node pairs	*Primary (P) WCHs*	*Links*	λ	*Spare (S) WCHs*	*Links*	λ
a b	1	*A*	λ_1	2	*EB*	λ_1
a c	3	*D*	λ_1	4	*EC*	λ_1
a d	5	*E*	λ_4	6	*AB*	λ_4
b c	7	*AD*	λ_2	8	*BC*	λ_3
b d	9	*B*	λ_2	10	*AE*	λ_3
c d	11	*C*	λ_2	12	*DE*	λ_2

5 COMPARISON OF WAVELENGTH ASSIGNMENT RESULTS

The wavelength assignment procedure was carried out for a core part of the European Optical Network, a case study developed within the European Commission project COST 239 "Ultra-high capacity optical transmission networks" (Figure 4). The network consists of 11 nodes. The total traffic requirements are fulfilled by 75 *WCH*s, each 10 Gbit/s. The path protection scenario is assumed.

The results are presented in Table 2. The comparison is done for *GA*, *SA*, and *GC* approaches, taking into account different numbers of *WCH*s per fiber N_λ.

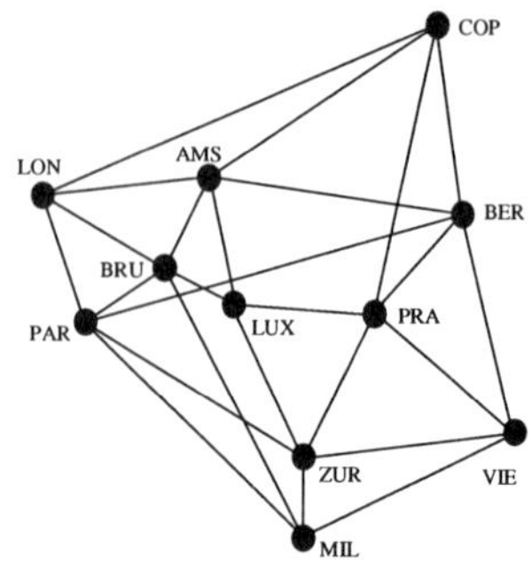

Figure 4 A core part of the European Optical Network – a case study COST 239

Table 2 Comparison of wavelength assignment results for the case study

	$N_\lambda = 4$			$N_\lambda = 8$		
	GA	*SA*	*GC*	*GA*	*SA*	*GC*
$\max_k\{F_k(i)\}$	12	14	10	10	12	6
$\Sigma_k F_k(i)$	178	190	158	134	150	92
$\Sigma_k w_k(i)$	500	502	472	542	504	482
UF	0.702	0.637	0.716	0.491	0.417	0.634
TFL (km)	91 400	97 560	84 820	71 710	79 360	49 780

6 CONCLUSION

Two general heuristic search techniques, *GA* and *SA*, were applied in solving the wavelength assignment problem for the case study. Both techniques inherently do not use any additional knowledge about the problem. On the other hand, in the proposed approach the wavelength assignment is transformed to the graph-coloring problem, by using a great amount of the knowledge about the case. The determinism introduced produced better optimization results in a shorter running time. An analysis of searching results shows that *SA* is more sensitive to the values of control parameters (cooling schedule) than *GA* (crossover and mutation probability). Careful adjustment of the control parameters and additional experiments could improve optimization results in *GA* and *SA*, but it seems that a hybrid approach, where first independence/disjointedness tests are performed and then *GA* or *SA* are used in graph coloring, with a goal to minimize the total fiber length in the network.

7 REFERENCES

Goldberg, D.E. (1989) Genetic Algorithms in Search, Optimization and Machine Learning, Addison Wesley.

Aarts, E. and Korst, J. (1990) Simulated Annealing and Boltzman Machines, John Wiley & Sons.

Inkret, R., Podnar, I., and Mikac, B. (1997) Minimization of Total Fiber Length in All-optical Networks Using Genetic Algorithm, *Proc. Conf. on Computers in Telecommunications* MIPRO 97, Opatija, Croatia, May 1997, pp. 31-36.

Welsh, D.J.A. and Powell, M.B. (1967) An upper bound for the chromatic number of a graph and its application to timetable problems, *Computing Journal*, **10,** pp. 85-96, 1967.

8 BIOGRAPHY

Branko Mikac was born in 1947, Gospic, Croatia. He received his B.Sc., M.Sc. and Ph.D. in electrical engineering from the Faculty of Electrical Engineering, University of Zagreb, Croatia, in 1970, 1980 and 1986, respectively. From 1974, he is with the Faculty. His research and lecture interests include transmission systems and reliability of telecommunication systems. He is a member of IEEE and Croatian Telecommunication Society. He is involved in project COST 239 "Ultra-high Capacity Optical Transmission Networks".

Robert Inkret was born in 1972, Zagreb, Croatia. He received his B.Sc. in electrical engineering from the Faculty of Electrical Engineering and Computing, University of Zagreb, in 1995. Since 1996 he has been an assistant at the Faculty, Department of Telecommunications. His research interests include digital transmission and system reliability analysis. He is involved in COST 239.

Francois Tillerot was born in 1969, Autun, France. He received the Diplome d'Ingenieur de l'Ecole Polytechnique, Palaiseau, France in 1991, the Diplome d'Ingenieur des Telecommunications and the Diplome d'Etudes Approfondies (DEA) in Optoelectronics from Telecom Paris, France in 1993. He joined the Centre National d'Etudes des Telecommunications (CNET), Lannion, France in 1993. His present research lies in the study of wavelength routing in all optical transport networks. He is involved in COST 239 and ACTS OPEN projects.

PART THREE

Optical technologies for local and access networks

An optically transparent ultra high speed LAN-ring employing OTDM

K. Bengi, G. Remšak, H.R. van As

Vienna University of Technology,

Institute of Communication Networks

Gusshausstrasse 25/388, A-1040 Vienna, Austria

Tel.: +43-1-58801-3518, Fax: +43-1-5870583,

email:kemal.bengi@tuwien.ac.at

Abstract

In this paper, an optically transparent 100Gbit/s ring-LAN architecture based on the slotted ring protocol and the OTDM technique is presented. Optical transparency is achieved by using optical packets of constant duration which consist of variable bit rate payloads and headers with fixed data rate. Thus, the considered LAN is capable of providing multiple services for nodes with different transmission speeds.

Keywords

photonic networks, ring-LAN, OTDM

1. INTRODUCTION

In the recent past, Local Area Networks (LANs) with various network topologies and media access methods have been proposed. Nowadays, LANs must be capable of integrating multiple services like multimedia, videoconferencing, fast database access, high-performance interprocessor communications and many other high- bandwidth applications. For that reason, future networks have to provide data rates in the upper Gbit/s range on instantaneous demand with high overall channel efficiency and minimal latency, which can only be realized by photonic networks.

Currently, the utilization of the optical fiber bandwidth in high-speed shared medium LANs is limited to a few Gbit/s due to the electronic bottleneck in processing the optical signals in the network nodes. In consequence, the nodes of a conventional local area network communicate at a single low bit rate with each other, since the handling of the data packets is presently performed electronically using optoelectronic conversion at each node. In order to achieve aggregate data rates up to Tbit/s ranges *optical* processing methods have to be applied, thereby removing the conversion between the electrical and optical domain in the optical signal path.

A very promising technique in this context is *optical time-division multiplexing (OTDM)*, which yields aggregate capacities beyond the reach of high-speed electronics. In general, OTDM (local area) networks provide interconnections with very high guaranteed bandwidth, low latency and broadcast capability resulting in bandwidth on demand with low access delay. In contrast to WDM networks, which divide fiber bandwidth into many optical wavebands carrying information at electronic rates (≈ 2.5-40Gbit/s), OTDM networks use short pulses to transmit at very high bit rates (> 100Gbit/s) in a single waveband.

The main advantages of OTDM networks are in general:

- interconnection of ultra high-speed nodes and networks
- fast transmission of very large data blocks
- switching of large traffic volume
- flexible access to the medium

Thus, OTDM networks are suitable for satisfying the demand for mixed high-speed traffic. There are several theoretical and experimental investigations on OTDM bus networks to date ([Ellis, 1994], [Lucek, 1996], [Barry, 1996]).

In this paper, an optically transparent 100Gbit/s ring-LAN architecture based on the slotted ring mechanism and employing the OTDM technique is presented. Optical transparency is achieved by using optical packets of constant duration which consist of variable bit rate payloads and headers with fixed data rate. Furthermore, special

OTDM nodes, at which multiple channels are multiplexed by OTDM, are integrated into the network.

2 NETWORK ARCHITECTURE

The proposed network serves nodes with different processing capabilities. While the optical medium operates at a line-rate of maximally 100Gbit/s, the user bandwidth requirements vary. The media access protocol of the LAN is based on the *slotted ring* technique with *destination release* and *spatial slot reuse*. Fixed-period slots carrying optical packets circulate permanently on the ring and can be empty or full. The ring nodes with traffic from attached stations can fill multiple slots in each ring revolution with their data packets. The choice of this protocol is due to its suitability for real-time traffic and for network transparency. Destination release is proposed since its use results in efficient channel utilization and an increase in ring capacity. In consequence, by allowing destination release and reuse of slots, the network throughput significantly exceeds the transmission rate of the medium. In contrast to alternative multiplexing methods such as WDM or bit-interleaved TDM ([Ellis, 1994]), the slotted TDM scheme enables sharing one fast medium by access nodes capable of operating at 100Gbit/s rates. Thus, we can design a network architecture which simultaneously provides guaranteed bandwidth and random access bandwidth on demand services. Consequently, the provision of multiple services to a heterogeneous user population becomes feasible.

Figure 1 demonstrates the basic principle of operation of the considered ring-LAN.

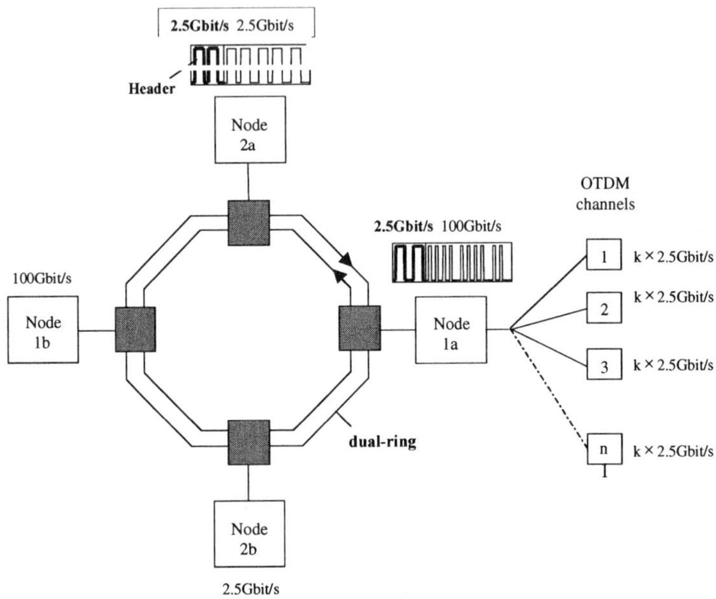

Figure 1 Basic configuration of the optical transparent ring-LAN.

As it can be seen in the figure, Node 1a has the capability of transmitting with an exemplary bit rate of 100Gbit/s by optical time-division multiplexing of various local channels. These channels operate at *k* multiples of the arbitrarily chosen standard data rate of 2.5Gbit/s. The header speed is uniformly maintained at 2.5Gbit/s across the entire network. Since Node 1b also represents an OTDM node equipped with OTDM demultiplexer and optical clock recovery facilities, it can properly process the corresponding 100Gbit/s signal sent from Node 1a. Similarly, the 2.5Gbit/s

conventional, i.e. only electronically operating, Node 2a can communicate with the corresponding Node 2b directly and straightforward. The data exchange between low bit rate nodes like Node 2a, 2b and OTDM nodes (Node 1a, 1b) is also easily feasible, since the optical clock associated with the OTDM demultiplexer may be adjusted to the appropriate repetition rate (in this case to 2.5Gbit/s) in order to extract the desired low bit rate signal. Naturally, the OTDM nodes have to know about the processing capabilities of their destination nodes, when they want to transmit packets to them.

The slots for the packets to transmit have a fixed duration and consist of a *payload* and a *header* section. Moreover, they are separated by *guard bands*. For example, a network with the length of 1km and consisting of 10 attached nodes and 40 slots yields a slot period of 125ns. The payload is to be transparent to bit rate and transmission format. The header contains source/destination addresses and control information (full/empty flag, monitor field, priority fields).

Due to the dual-ring configuration, the shortest path among the two fiber loops is chosen for effectively transmitting the optical packets to the specific destination.

The desired optical transparency is achieved by the provision of bit rate variable payloads and a header with fixed bit rate. Consequently, the optical packets consist of mixed rate data streams ([Cai, 1994]). Since the attached nodes operate at different speeds and the generation of optical signals with very high bit rates (20– 100Gbit/s) is performed by OTDM, we have to differentiate between the conventional electronic and the optical processing of the optical packets. Therefore, optically transparent photonic packet switching is to be used allowing payload bit rates to be independent of the header's data rate. Thus, only the header of each packet needs to be read by all the ring nodes, whereas the payload does not need to be perceived by all nodes. The bit rate of the payload is determined by the transmitting node, based on the capabilities of the corresponding receiving nodes. For that reason, different network nodes are capable of operating at different bit rates simultaneously. In consequence, this ensures that higher-speed nodes may be added to the network at a later date without requiring any changes in the lower-speed nodes currently operating in the network. On the other hand, this approach requires that each node knows about the bit rate capabilities of the other network nodes. This can be achieved either by using look-up tables for the achievable transmission speeds of the individual nodes on the ring or by polling each node before establishing a connection to that node. The lowest bit rate employed within the network could be 2.5Gbit/s (corresponding to the STM-16 format of SDH systems). In order to enable nodes to operate at very high bit rates like 100Gbit/s, OTDM is to be used. This can be done by multiplexing various channels of a specific node with respect to the processable bit rate of the destination node. If the destination nodes are not designed for operating at very high data rates, the corresponding OTDM source node may switch to an *electronical* transmission unit.

3 NODE DESIGN

In order to provide optical transparency, the nodes detect only the headers operating at a fixed low bit rate and the payload of a packet is only read when it is destined to a specific node. Thus, an inexpensive node with a lower transmission speed does not need to process the optical time-division multiplexed signal from an expensive node operating, for example, at 100Gbit/s. In the proposed network architecture, there are two main node configurations to be considered. Firstly, nodes operating at low bit rates up to 20Gbit/s and secondly the nodes using OTDM to achieve ultra high bit rates like 100Gbit/s, for which the proposed network is mainly designed.

The principle of the node's operation, which extends the scheme in [Cai, 1994] is demonstrated in Figure 2, where specifically the configuration for the OTDM node is depicted.

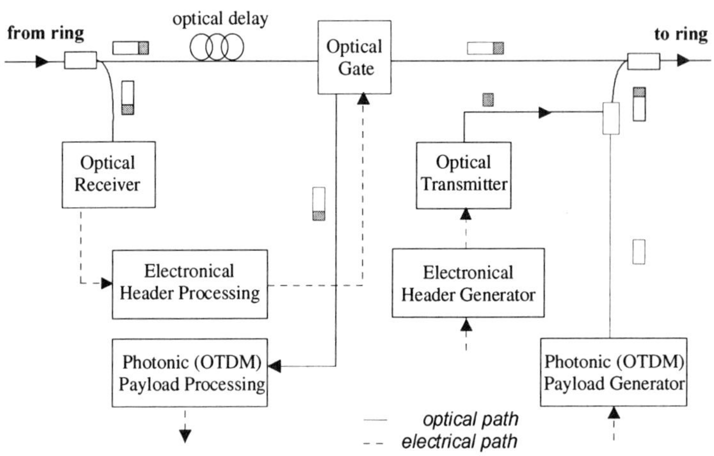

Figure 2 OTDM node configuration.

The block diagram for the low bit rate nodes operating electronically can be derived directly from the OTDM case. Let us consider the functionality of the high-speed node.

The *input fiber coupler* couples a fraction (commonly 10/90) of the light from the incoming optical slot. An *optical delay line* delays the slot while the header is received and processed. For the dimensioning of the optical delay line also the delays at the transmitter side of the node has to be taken into account. The headers (e.g. 2.5Gbit/s) of all slots that pass the node are detected by the *optical receiver* and processed by the *electronical header processing* block. Then, the electronic control unit triggers the further operations of the node components depending on whether the slots are addressed to the corresponding node or not. Particularly, it triggers also the transmitter part, which is not explicitly indicated in Figure 2.

In general, three possible situations could occur:

(1) If an incoming slot is not addressed to the node, it simply propagates through the node (i.e the optical gate is set to the "*on*" state).

(2) If the destination address of an incoming slot matches the address of the node, the *payload processing* block reads the payload data (using an passive coupler integrated in the *optical gate* block) and removes it from the ring according to the destination release rule, while the optical gate is set to the "*off*" state, thus blocking the transmission of the slot. Simultaneously, the node's transmitter section generates either an empty or a full slot depending on the existence of new data to be transmitted. The generation of an empty slot is straightforward since it does not require photonic operations. A new filled slot has to be created by independently generating the header (via the *electronical header generator*) and the payload. In order to obtain the payload, various local channels of the OTDM node have to be optically multiplexed via an *OTDM multiplexer* to achieve the desired very high transmission speeds. Then, the resulting OTDM payload signal is appended to the header. After combining header and high-speed payload, the new slot is put onto the ring via the *output fiber coupler*. Concerning the receiver part of the node, it has to be noted that the *payload processing* block consists of typical OTDM components such as the *OTDM demultiplexer* and *the optical clock recovery unit.* In the case of the low bit rate nodes, the payload is processed electronically. Thus, the node requires an additional O/E conversion between the *optical gate* and the *payload processing unit*.

(3) If an incoming slot is empty, indicated by an empty flag or by continuous wave light, the node inserts a new packet onto the ring by using the *OTDM multiplexer*, while the transmission of the empty slot is blocked through the optical gate. In the case of low bit rate nodes and indicating an empty slot by a cw-signal there is also the possibility to modulate the cw-signal by the optical gate to insert a new packet.

As optical gate within the node a *semiconductor optical amplifier (SOA)* could be employed to gate or block packets, modulate packets and providing optical gain to overcome losses in the system.

4 CONCLUSION

In order to achieve ultra high-speed LANs fulfilling the requirements of high-bandwidth applications, an optically transparent ring network with a maximum line-rate of 100Gbit/s employing the OTDM technique has been presented. The considered LAN is based on the slotted ring mechanism with destination release for its advantages in terms of optical transparency, bandwidth efficiency and transmission delays. Due to the concept of bit rate variable payloads, expensive nodes with high bit rate capabilities may coexist with low-cost nodes operating at low data rates. The main advantage of the proposed architecture consists in the possibility of supplying a heterogeneous user population with guaranteed bandwidth and bandwidth on demand services. Currently, a specific MAC (Medium Access Control) protocol for the introduced LAN architecture is being developed in order to guarantee fairness for the attached nodes of the ring-LAN.

5 REFERENCES

Ellis A.D., Widdowson T. et. al. (1994) Three-node, 40Gbit/s OTDM network experiment using electro-optic switches. *Electronics Letters,* **16**, 1333–34.

Lucek J.K., Gunning P. et. al. (1996) 40Gbit/s Optical TDMA LAN. *ECOC'96(Oslo) – Proceedings ,* 5.45–5.48.

Gunning P., Lucek J.K. et. al. (1997) 40Gbit/s Optical-TDMA LAN Over 300 Metres Installed Blown Fibre. *ECOC'97(Edinburgh) – Proceedings*, 61–64 .

Barry R.A., Chan V. et. al. (1996) All-Optical Network Consortium – Ultrafast TDM Networks. *IEEE Journal on Selected Areas in Communication,* **5**, 999–1013.

Cai Y. et. al. (1994) Demonstration of Photonic Packet-Switched Ring Network with Optically Transparent Nodes. *IEEE Photonics Technology Letters,* **9**, 1139–41.

Chan C.K. et. al. (1995) Experimental demonstration of a high-speed all-optical tunable-channel multi-access (TCMA) network. *Globecom'95,* 1761–65 .

6 BIOGRAPHY

Kemal Bengi received his Dipl.-Ing. degree in Electrical Engineering from the *University of Erlangen (Germany)* in 1997. Since May 1997 he is research assistant working towards his Ph.D. degree at the *Institute of Communication Networks, Vienna University of Technology*. His research interests include *photonic networks, MAC protocol design* and *performance evaluation.*

Günter Remšak received his Dipl.-Ing. degree in Technical Physics from the *Vienna University of Technology* in 1996. Since March 1997 he is research assistant working towards his Ph.D. degree at the *Institute of Communication Networks, Vienna University of Technology*. His research interests include *photonic networks, optical transmission* and *performance evaluation.*

Harmen R. van As received his Dipl.-Ing. degree in Electrical Engineering from the *ETH Zurich* and the Ph.D. degree from the *University of Siegen (Germany)*. Since March 1996 he is full-professor and head of the *Institute of Communication Networks, Vienna University of Technology.*

Efficient Multi-Hop Scheduling Algorithms for All-Optical WDM/TDM Broadcast-and-Select Networks with Arbitrary Transceiver Tuning Latencies

M. Ajmone Marsan, A. Bianco, E. Leonardi, F. Neri, A. Nucci
Dipartimento di Elettronica, Politecnico di Torino
Corso Duca degli Abruzzi 24, 10129 Torino, Italy
Phone: +39 11 564 4000, Fax: +39 11 564 4099
E-mail: {ajmone,bianco,leonardi,neri}@polito.it

Abstract

We focus on optical slotted WDM/TDM broadcast-and-select networks with N wavelengths and N nodes. Nodes are equipped with one tunable transmitter with non-negligible tuning times and one fixed receiver. We propose a novel scheduling algorithm, exploiting the multi-hop approach, to shorten the duration of the transmission scheduling period. A simulation based analysis of the algorithm shows that it leads to significant performance improvements with respect to previous single-hop proposals. We also describe a single-hop scheduling algorithm that performs slightly better than previous proposals

Keywords

Optical Networks, Transmission Scheduling, Multihop Scheduling, Tuning Latency

INTRODUCTION

All-optical broadcast-and-select networks that use WDM (wavelength division multiplexing) as well as TDM (time division multiplexing) can provide transparent support to several simultaneous very-high-speed end-to-end communications, provided that adequate protocols are defined to dynamically share the available network resources.

A number of access protocol proposals for all-optical WDM/TDM broadcast-and-select networks appeared in the literature, some based on random access approaches (see [1] and references therein), some based on a deterministic schedule of the transmissions from the different traffic sources [2, 3, 4] in circuit switching scenarios; in this paper we discuss novel scheduling protocols of the latter type.

We assume that the all-optical WDM/TDM broadcast-and-select network is based on a broadcast topology (e.g. a star) in which W wavelengths are available for node-to-node communications, and that nodes are equipped with **one** full-duplex transceiver; hence, each node can be source and destination of at most one data flow at any given time. By tuning the nodes transceivers and by dynamically allocating the W available wavelengths, full network connectivity is achieved. Transmission channels are assumed to be slotted and synchronized; each slot can accommodate the transmission of one packet.

The time required to tune the node transceivers is *not* assumed to be negligible with respect to the packet transmission times; in fact, with the presently available optical components, the tuning latency can be equal to several packet transmission times [5, 6]. Access protocols that require a separate tuning action for each packet transmission yield very limited throughput under these conditions. As a consequence, it is necessary to devise simple (hence fast) algorithms to contiguously schedule transmissions on the same wavelength, so that transceiver tuning times do not significantly affect the network throughput.

The problem of finding an optimum transmission schedule in presence of significant tuning latencies was already extensively analyzed [7, 8, 9]. It can be formalized in the following way.

> GIVEN a requested traffic matrix R, whose elements r_{sd} are the numbers of packets that must be transmitted from any node s to any node d in a scheduling period, FIND a time/wavelength assignment that guarantees the delivery of the requested traffic, while MINIMIZING the time necessary to accommodate all transmissions, SUBJECT TO technological constraints.

In [7] the single-hop scheduling problem was shown to be NP-hard. Several heuristic approaches for the determination of good schedules were proposed by other authors [8, 9].

Most of the proposals that appeared in the literature aimed at the minimization of the scheduling period duration assuming that all packets must be transmitted in single-hop fashion (i.e., packets are directly transferred from their source to their destination with just one transmission). This may not necessarily be the best approach when tuning latencies are long.

In this paper, we present a novel heuristic approach for the identification of good single-hop schedules, which yields slightly larger throughput than the schedules resulting from previous proposals. Moreover, we show that significant throughput gains can be obtained by transmitting selected packet flows in multi-hop fashion, i.e., by routing them through intermediate nodes. Since we aim at throughput maximization, we do not optimize delays, and we assume that, in the case of multi-hop transmissions, the successive hops that a packet must traverse are scheduled in successive scheduling frames.

The problem of defining an optimal schedule (i.e., a schedule that maximizes the system throughput) when some traffic flows can be transmitted in multi-hop fashion, will be formalized as an Integer Linear Programming (ILP) problem, hence shown to be NP-hard.

To reduce complexity, we propose in this paper a heuristic approach to the multi-hop scheduling problem solution, based on two separate steps. In the first step, the set of traffic flows that are conveniently transmitted in either single-hop or multi-hop fashion is determined; for each multi-hop flow, the flow routing, i.e., the set of intermediate nodes traversed by the flow, is also defined. As a result of this first step, the single-hop traffic matrix S is derived from R: each element s_{ij} of S is equal to the sum of the number of packets of the traffic flows that, according to the solution found, are to be transmitted in single-hop fashion from node i to node j. In the second step, a single-hop scheduling algorithm is run in order to define the time/wavelength schedule for packet transmissions. The algorithm adopted in this second step represents itself an improvement over previously proposed algorithms.

THE MULTI-HOP SCHEDULING PROBLEM

We consider a network with N nodes, each node being provided with one tunable transmitter and one fixed receiver. We denote by W the number of available wavelengths, and we assume $W = N$, so that each wavelength leads to a different destination. The transmitters tuning latency, i.e. the time required to tune from a wavelength to a different one, is taken to be T slots. Thus, two transmissions from the same source at different wavelengths must be separated by at least T empty slots.

Although we consider in this paper the case of transmitter tunability, both the formalization of the problem, and the proposed heuristic, can easily be extended to the case of receiver tunability, and to a more general context in which both transmitters and receivers are tunable with significant latencies; similar extensions can be devised if the number of wavelengths is not equal to the number of nodes.

Problem Formalization

Define the binary variables v^{ij} that describe the logical topology of the network, i.e. the pattern of logical links in the network, where i and j refer to node indexes. A logical link $i \rightarrow j$ exists if packets are transmitted in single-hop fashion from node i to node j:

$$v^{ij} = \begin{cases} 1 & \text{if a logical link } i \rightarrow j \text{ exists} \\ 0 & \text{otherwise} \end{cases}$$

Define the binary variables b_{ij}^{sd} that describe the routing on the logical topology, where i, j, s, and d all refer to node indexes, such that:

$$b_{ij}^{sd} = \begin{cases} 1 & \text{if the traffic flow } s \Rightarrow d \text{ is routed through link } i \rightarrow j \\ 0 & \text{otherwise} \end{cases}$$

If we assume that the whole traffic flow from node s to node d follows the same route (this constraint guarantees the sequential delivery of transmitted packets to their destinations), the variables v^{ij} and b_{ij}^{sd} must satisfy the following constraints, deriving from such assumption, from the fact that traffic flows must travel over existing logical links, and from the flow conservation at each intermediate node:

$$\sum_i b_{id}^{sd} \leq 1 \tag{1}$$

$$\sum_j b_{sj}^{sd} \leq 1 \tag{2}$$

$$b_{ij}^{sd} \leq v^{ij} \tag{3}$$

$$\sum_i b_{ij}^{sd} = \sum_k b_{jk}^{sd} \qquad \text{if } j \neq s, \;\; j \neq d \tag{4}$$

Define the binary variables t_l^{ij}, that describe the transmission activity, where superscripts i and j refer to node indexes, and subscript l refers to time slot labels, such that:

$$t_l^{ij} = \begin{cases} 1 & \text{if node } i \text{ transmits to node } j \text{ in time slot } l \\ 0 & \text{otherwise} \end{cases}$$

The variables t_l^{ij} must satisfy the following constraints, deriving from the fact that nodes can transmit and receive no more than one packet in each slot, and from the tuning time constraint:

$$\sum_i t_l^{ij} \leq 1 \tag{5}$$

$$\sum_j t_l^{ij} \leq 1 \tag{6}$$

$$t_l^{ij_1} + t_{l+k}^{ij_2} \leq 1 \quad \forall k = 1, \cdots, T \quad j_1 \neq j_2 \tag{7}$$

If all requests in R must be satisfied, the variables t_l^{ij} and b_{jk}^{sd} are related by:

$$\sum_l t_l^{ij} = \sum_s \sum_d r_{sd} b_{ij}^{sd} \tag{8}$$

where r_{sd}, the elements of the request matrix R, represent the number of packets that must be transferred from node s to node d during the considered scheduling period.

Finally, we introduce the binary variables e_l^i to identify an upper bound to the scheduling frame duration. The variables e_l^i, where i is a node index and l the time slot label in a scheduling frame, must meet the following constraints:

$$e_l^i \geq e_{l+1}^i \tag{9}$$

$$e_l^i \geq t_l^{ij} \tag{10}$$

$$e_{l+T-k+1}^i = 1 \quad \text{if } \exists j_1 \neq j_2 \text{ and } \exists k,\ 1 \leq k \leq T : t_k^{ij_1} + t_l^{ij_2} > 1 \tag{11}$$

Note that inequalities (9) and (10) force $e_l^i = 1$ for all the slots from the beginning of the scheduling period to the end of the last transmission by node i. Inequality (11) forces $e_l^i = 1$ for an additional tuning time, if not enough idle slots are available at the end of the scheduling period or at the beginning of the following one.

With the above definitions, the criterion for the optimization of the multi-hop schedule is the minimization of the transmission period F satisfying:

$$F \geq \sum_l e_l^i \quad \forall i \tag{12}$$

This minimization requires the solution of an ILP problem and, as a consequence, it is NP-hard.

The Heuristic Algorithm

Given the complexity of the multi-hop scheduling problem, we devised a heuristic algorithm which, given a traffic request matrix R, efficiently schedules packet transmissions on an all-optical WDM/TDM broadcast-and-select networks.

Recall that the elements r_{sd} of the traffic request matrix R represent the number of packets that must be transmitted from source s to destination d in a scheduling period. If $L(R)_i$ is the minimum duration of the activity of node i

$$L(R)_i = \max\left(\sum_j r_{ij} + K_i T, \sum_k r_{ki}\right) \tag{13}$$

where K_i is the number of destination nodes to which node i must transmit at least one packet, the length L_{sh} of the single-hop scheduling period (i.e. the period necessary to accommodate all transmissions in R in single-hop fashion) must satisfy:

$$L_{sh} \geq L(R) = \max_i L(R)_i \tag{14}$$

Since we assume that the successive hops of a multi-hop transmission occur in successive scheduling periods, we can split in two separate steps the problem of defining a packet schedule that minimizes the scheduling period duration, hence maximizes the throughput, while allowing multi-hop transmissions.

The first step aims at the design of a logical topology through which packets must be routed from sources to the destinations; i.e., at the definition of the set of single-hop transmissions as well as the routing for multi-hop transmissions. During the second step, the single-hop scheduling problem is solved starting from a derived traffic matrix S, whose elements s_{ij} are equal to the number of packets that are to be transmitted in single-hop fashion from node i to node j, possibly comprising multi-hop packet flows, according to the solution found in the first step. A novel heuristic single-hop scheduling procedure is proposed, that performs slightly better than previous proposals [8, 9], and that allows the achievement of scheduling period lengths very close to $L(R)$ for non-negligible values of the tuning latency.

We describe the logical topology construction process with an algorithm. Let $S^{(n)}$ represent the single-hop scheduling matrix at step n; define, for each node i, the length of the transmission schedule at step n.

$$L_i^{(n)} = \sum_j s_{ij}^{(n)} + K_i T \tag{15}$$

The logical topology design algorithm comprises the following steps.

1. Initialize $S^{(0)}$ to R: all connections are enabled (they take place in single-hop fashion).

2. Among all sources of enabled connections, choose node i for which $L_i^{(n)}$ is maximum.

3. Among all destinations of enabled connections whose source node is i, find the destination node j for which $s_{ij}^{(n)}$ is minimum.

4. Find a *pivot* node k so that $s_{ik}^{(n)} > 0$ and $s_{kj}^{(n)} > 0$, through which the flow $i \Rightarrow j$ could be routed in multi-hop fashion; if many such nodes exist, choose the one for which $L(S^{(n)})_k$ is minimum.

5. If a pivot node has been found, compute the temporary matrix S^* as follows: first $S^* = S^{(n)}$, then $s^*_{ij} = 0$, $s^*_{ik} = s^{(n)}_{ik} + s^{(n)}_{ij}$, and $s^*_{kj} = s^{(n)}_{kj} + s^{(n)}_{ij}$.

6. If $L(S^*) < L(S^{(n)})$, then the rerouting of connection (i, j) takes place: $S^{(n+1)} = S^*$.

7. Connection $i \rightarrow j$ is disabled. If any enabled connection is left, go to Step 2; otherwise $S = S^{(n)}$.

In the second step, the single-hop scheduling strategy is solved using the following algorithm.
For each source-destination pair (i, j), define two functions: $Source(i, j)$ which returns i and $Dest(i, j)$ which returns j. Moreover, let $\mathcal{A}$ denote the set of assigned connections, i.e. the set of source-destination pairs (i, j) corresponding to packet flows whose transmission has already been scheduled; $\mathcal{A}$ is initialized to the empty set $\emptyset$. For each source node i, define the set $\mathcal{D}(i)$ of destinations corresponding to already scheduled transmissions; $\mathcal{D}(i)$ is initialized to $\emptyset$ for each node i. Let t_s denote the system time and initialize $t_s^{(0)} = 0$. At the algorithm start, all sources and destinations are enabled. The algorithm consists of the following steps.

1. Use a Maximum Weighted Matching (MWM) algorithm (see [10]) over the requests issued by enabled source nodes for transmission toward the enabled destinations, in order to find the switching configuration, i.e. the maximal set of concurrently active transmissions; for each pair (i, j) corresponding to transmissions selected by the MWM algorithm, put $Dest(i, j)$ in $\mathcal{D}(Source(i, j))$, put (i, j) in $\mathcal{A}$, and disable the corresponding source and destination nodes.

2. Update the traffic matrix S by subtracting all the entries corresponding to transmissions on connections that were just added to

$\mathcal{A}$.

3. For each source node i, evaluate $t_i^{(n)}$, the minimum time at which the node can schedule a transmission toward a destination not belonging to $\mathcal{D}(i)$; let $t^{(n)}$ be the minimum value of $t_i^{(n)}$ greater than t_s. Update $t_s^{(n+1)} = t^{(n)}$. Enable the nodes for which $t_i^{(n)}$ is not greater than $t_s^{(n+1)}$.

4. Enable the available destinations in slot $t_s^{(n+1)}$, i.e. the destinations that are not receiving any packet in slot $t_s^{(n+1)}$.

5. If matrix S is null, the algorithm terminates. Otherwise, consider S_{en}, the submatrix of S referring to enabled source and destination nodes; if S_{en} is not null, go back to step 1. If S_{en} is null, go back to step 3.

Since the algorithm is based on an incremental application of the MWM algorithm, we shall name it IMWM (Incremental MWM).

NUMERICAL RESULTS

In order to provide a first indication about the effectiveness of the proposed multi-hop scheduling approach in all-optical WDM/TDM broadcast-and-select networks, we present some numerical results obtained by running the proposed single-hop and multi-hop scheduling algorithms with an approximate implementation of the MWM algorithm.

We consider networks with a number of nodes N and a number of wavelengths W equal to 32. One hundred different traffic matrices R are randomly generated. The elements of matrix R correspond to individual requests from source nodes, and are taken to be geometrically distributed with mean equal to 10. The elements along the main diagonal are null, meaning that nodes do not transmit to themselves.

The results of the scheduling algorithms are plotted in Figure 1 as curves of scheduling period duration (averaged over the 100 randomly gener-

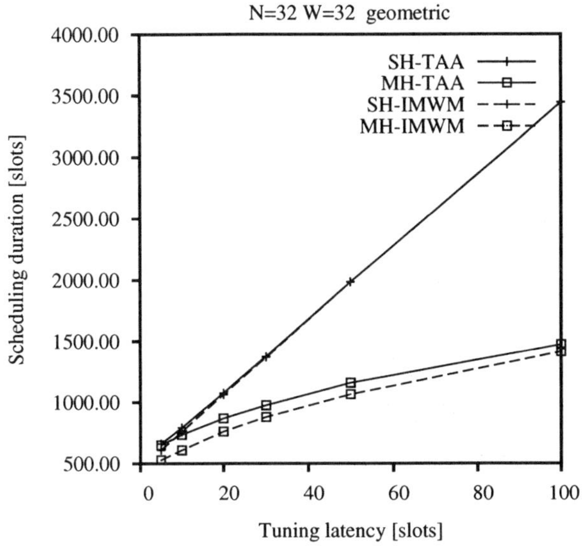

Figure 1: Scheduling period duration as a function of the number of slots necessary to tune node transmitters for the single-hop and multi-hop TAA, and single-hop and multi-hop IMWM algorithms

ated matrices R) versus the number T of slots necessary to tune the transmitters. As regards the single-hop scheduling algorithm, we consider both IMWM, newly proposed in this paper, and TAA, originally proposed in [9]. For the multi-hop scheduling, we always consider our heuristic approach for the design of the logical topology, but again we consider either IMWM or TAA for the second step of the procedure.

The curves in Figure 1 show that the gain obtained with the multi-hop scheduling algorithm can be quite remarkable, specially with long tuning times. Instead, the differences between the two considered single-hop scheduling algorithms do not appear to be significative, even if they slightly favor IMWM over TAA.

Observe that, although not previously mentioned, we can restrict the algorithm to schedule packets with a limited maximum number of hops, aiming at reduced packet delays. In Figure 2 we show the curves of

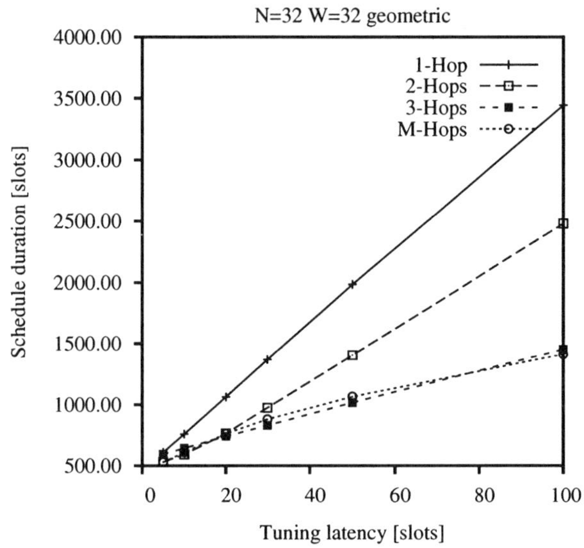

Figure 2: Scheduling period duration as a function of the number of slots necessary to tune node transmitters for the single-hop, 2-hops, 3-hops and multi-hop IMWM algorithms

the average scheduling period duration versus the tuning time in slots, for the case of a maximum number of hops equal to 1 (single-hop), 2, 3, or unlimited (multi-hop). We can observe that the improvements are significant from 1 to 2 hops and from 2 to 3 hops, but there is no evident gain when no restriction on the maximum number of hops is imposed. Actually, the multi-hop case for some tuning delays even produces worse results than the 3-hop case. This is due to the fact that when the number of hops grows, the resulting matrix S is more difficult to schedule in single-hop mode.

CONCLUSIONS

We have considered all-optical WDM/TDM broadcast-and-select networks with N nodes and $W = N$ slotted and synchronized transmission channels. Network nodes have one fixed receiver and one tunable transmitter with non-negligible tuning times.

For these systems, we proposed to adopt the multi-hop scheduling of WDM/ TDM packet transmissions. We first formally defined the multi-hop scheduling problem as an ILP (Integer Linear Programming) problem. To reduce the complexity, we then described a heuristic approach to the problem, which runs through two steps, called the logical topology design and the single-hop scheduling. An algorithm which provides a sub-optimal solution to the logical topology design problem was proposed. We have further introduced an algorithm for the single-hop scheduling problem which provides minor improvements over previously proposed algorithms.

Our results show that the adoption of multi-hop scheduling algorithms provides important throughput advantages over single-hop scheduling. We also showed that limiting the maximum number of intermediate nodes to 2 leads to a good compromise in terms of throughput vs. delay, with respect to single-hop and unbounded multi-hop scheduling algorithms.

REFERENCES

[1] B.Mukherjee, "WDM Based Local Lightwave Networks – Part I: Single Hop Systems", IEEE Network Magazine, vol.6, pp.12-27, May 1992.

[2] J.C.Lu, L.Kleinrock, "A Wavelength Division Multiple Access Protocol for High-Speed Local Area Networks with Passive Star Topology", Performance Evaluation Journal, vol.16, pp.223-239, Nov.1992.

[3] D.Guo, Y.Yemini, Z.Zhang, "Scalable High Speed Protocols for WDM Star Networks", in Proc. of IEEE INFOCOM'94, Toronto, Ontario, Canada, June 1994.

[4] F.Jia, B.Mukherjee, J.Iness, S.Ojha, "Variable Length Message Scheduling Algorithms for a WDM Based Local Lightwave Network", in Proc. of IEEE INFOCOM'94, Toronto, Ontario, Canada, June 1994.

[5] A.Sneh, K.M.Johnson, "High-Speed Tunable Liquid Crystal Optical Filter for WDM Systems", in Proc. of IEEE/LEOS'94, Summer Topical Meeting on Optical Networks and their Enabling Technologies, Lake Tahoe, NV, USA, July 1994.

[6] D.A.Smith et al., "Evolution of Acousto-Optic Wavelength Routing Switch", IEEE/OSA Journal of Lightwave Technology, Special Issue on Multiwavelength Optical Technology and Networks vol.14, n.6, pp.1005-1119, June 1996.

[7] I.S.Gopal, C.K.Wong, "Minimizing the Number of Switchings in a SS/TDMA System", IEEE Transactions on Communications, vol.COM-33, n.6, pp.497-501, June 1985.

[8] A.Ganz, Y.Gao, "A Time-Wavelength Assignment Algorithm For a WDM Star Network", in Proc. of IEEE INFOCOM'92, Florence, Italy, May 1992.

[9] M.S.Borella, B.Mukherjee, "Efficient Scheduling of Nonuniform Packet Traffic in a WDM/TDM Local Lightwave Network with Arbitrary Transceiver Tuning Latencies", IEEE Journal on Selected Areas of Communications, vol.SAC-14, n.5, pp.923-934, June 1996.

[10] N.McKeown, V.Anantharam, J.Walrand, "Achieving 100% Throughput in an Input-Queued Switch", in Proc. of IEEE INFOCOM'96, San Francisco, CA, USA, March 1996.

AUTHORS' BIOGRAPHIES

Marco Ajmone Marsan is a Full Professor at the Dipartimento di Elettronica of Politecnico di Torino, in Italy. He was born in Torino, Italy, in 1951. He holds a Dr.Ing. degree in Electronic Engineering from Politecnico di Torino, and a Master of Science from the University of California, Los Angeles. Since november 1975 to october 1987 he was at the Dipartimento Di Elettronica of Politecnico di Torino, first as a Researcher, then as an Associate Professor. Since November 1987 to October 1990 he was a Full Professor at the Computer Science Department of the University of Milan, in Italy. During the summers of 1980 and 1981 he was with the Research in Distributed Processing Group, Computer Science Department, UCLA. He has coauthored over 150 journal and conference papers in the areas of Communications and Computer Science, as well as the two books "Performance Models of Multiprocessor Systems" published by the MIT Press, and "Modelling with Generalized Stochastic Petri Nets" published by John Wiley. He received the best paper award at the Third International Conference on Distributed Computing Systems in Miami, Fla., in 1982. His current interests are in the fields of performance evaluation of data communication and computer systems, communication networks and queueing theory. M. Ajmone Marsan is a Senior Member of IEEE.

Andrea Bianco is an Assistant Professor at the Dipartimento di Elettronica of Politecnico di Torino, Italy, since December 1994. He holds a Dr.Ing. degree in Electronics Engineering and a Ph.D. in Telecommunications Engineering both from Politecnico di Torino. His current research interests are in the field of access protocols for all-optical networks and performance analysis of ATM networks.

Emilio Leonardi received the Dr.Ing. and Ph.D. degrees in Electronics Engineering from Politecnico di Torino in 1991 and 1996, respectively. From 1994 to 1995 he was a visiting scholar at the Computer Science Department of the University of California in Los Angeles (UCLA), working on deadlock-free routing and flow control techniques for asynchronous high speed wormhole routing networks. His current reasearch interests are in the fields of performance evaluation of high-speed and all-optical networks, and of queuing theory.

Fabio Neri is an Associate Professor at the Dipartimento di Elettronica of Politecnico di Torino, in Italy. He was born in Novara, Italy, in 1958. He received his Dr.Ing. and Ph.D. degree in Electronics Engineering from Politecnico di Torino in 1981 and 1987, respectively. From 1991 to 1992 he was with the Information Engineering Department at University of Parma, Parma, Italy, as an Associate Professor. From 1982 to 1983 he was a visiting scholar at George Washington University in Washington, DC. In the summer of 1995 he was visiting researcher at the Computer Science Department of the University of California in Los Angeles (UCLA). His research interests are in the fields of performance evaluation of communication networks, high-speed and all-optical networks, discrete event simulation, and queuing theory.

Antonio Nucci was born in Lecce, Italy, in 1974. He currently is working towards his Dr.Ing. degree in Electronics Engineering at Politecnico di Torino. His thesis regards scheduling algorithms and access protocols for optical networks.

Analysis of Wireless Broadband Systems employing Optical Fiber Links

Authors: M. Pallini, R. Sabella

Ericsson Telecomunicazioni, Rome, Italy

e-mail: sabella@RD.tei.ericsson.se

tel: 06-72583046 fax: 06-72583940

Abstract

In this paper we analyze the performance of fiber optic links for wireless broadband communication systems. Non-linear distortions arising from the laser source and noise contributions coming out from both transmitter and receiver are taken into account. With respect to the previous literature, a complete analysis of the optical link performance has been accomplished.

Keywords

FTTA system, millimeter-wave band, sub-carrier multiplexing, non-linear distortions, clipping, chirping

1 INTODUCTION

New subscriber distribution systems, called FTTA (Fiber To The Area), uses millimeter-wave radio band to connect subscribers with base station, and optical fiber to connect base station with control station, in order to obtain broadband transmission. In these systems, rainfall attenuation is the main source of signal degradation in the radio link, while the optical link is affected by several causes such as clipping due to the threshold characteristics of the laser source (assuming direct modulation that is the cheapest solution), the intermingling of laser chirping and fiber dispersion, and the signal distortion due to laser resonance frequency. While sufficient literature is available to characterize the radio link, an analysis of the simultaneous effect of the impairment occurring in the optical link of this type of system still has not been investigated. The aim of this paper is to present some figures about such effects on the system performance.

2 THE FTTA SYSTEM

The basic concept of an FTTA system is depicted in fig. 1 [1].
In this system each zone radius is 300 in order to enhance the spectrum utilization efficiency and to reduce transmitter power of both subscribers and base station. Base station receives RF signals from subscribers and converts them in optical signal by means of SCM (Subscriber Multiplexing) to transmit the control station through optical fiber. Then, the control station converts optical signal into RF signals, demodulates these signals, assigns each signal to channel and connects it to public switched networks.
The transmission, in this kind of system, can be schematized in a radio link and in an optical link.

The use of radio communication system with millimeter-wave is quite efficient in spectrum utilization, since it can cover small service zone with line-of-sight transmission and little co-channel interference. This radio transmission system is flexible for changing number of subscribers and it is easy to apply for high capacity mobile communications. On the other hand, the use of millimeter-wave band is limited by rainfall attenuation. In fact, if the rainfall induced attenuation exceeds a given threshold, outage occurs on radio channel.

The choice of carrier frequency for a radio system is usually dictated by its bandwidth and by the wave propagation characteristics (related to the extension of the coverage area), although antennas can pose restrictions as well (because of their size compared of the required gain). Thus, the study of wave propagation is an important task when developing a wireless systems. In this paper, we assume the use of 50 GHz as carrier frequency-band., for the radio transmission between subscriber and base station.
At this frequency band, modeling of wave propagation can be done on the basis of geometrical optics by using ray theory; at the millimeter wave band in particular, the diffraction phenomenon can be neglected the sum of the direct ray with reflected ones is enough to describe the propagation channel behavior with a reasonable degree of accuracy [2].

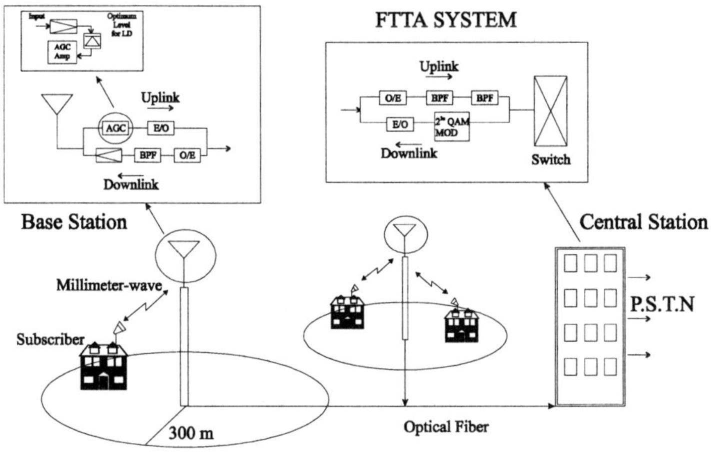

Figure 1. The concept of a Fiber-To-The-Air (FTTA) system

Definitely, the crucial parameter that has to be considered is the permissible attenuation due to rainfall, Za (dB). This can be evaluated according to the following relationship:

$$Z_a = P_T + G_R + G_T - L_f - KTBF - \gamma_{QAM} \qquad (1)$$

where the terms on the right hand side represent the transmission power from subscriber, the subscriber antenna gain, the base station antenna gain, the free-space loss, the noise from the receiver, and the required carrier-to-noise ratio (CNR), respectively.

In particular a relation exists between the required CNR values relating to 2^{2n} QAM signals transmission and to QPSK signal transmission:

$$\gamma_{\mathrm{QAM}} = \frac{1}{3}(2^{2n} - 1)\,\gamma_{\mathrm{QPSK}}\,. \qquad (2)$$

where γ_{QPSK} is the required CNR when QPSK is used. For instance, the value of γ_{QPSK} which guarantees that bit error rate is less than 10^{-6} is 13.8 dB [1].

The percentage of time in which the attenuation of the radio link exceeds the permissible attenuation Za represents the system outage. The evaluation of the outage probability can be carried out by evaluating such a percentage of outage time. Of course this value depends on the geographic location of the systems. Anyway this calculation is above the scope of the present paper.

3 ANALYSIS OF THE OPTICAL FIBER LINK

The radio-frequency (RF) signal transmitted from subscriber is converted into an optical signal, in the base station, by using a direct intensity modulation of a laser diode (LD). At the end the optical signal is transmitted to control station where it is converted into RF signals by a PIN photo-diode (PIN-PD).

The dynamics of a single mode laser diode can be modeled by the laser rate equations [3], which describe the interactions among photon density p(t), carrier density n(t), and an equation describing the phase of the electric field $\phi(t)$ behavior:

$$\frac{dp(t)}{dt} = \Gamma G(t)\left[n(t) - n_0\right]p(t) - \frac{p(t)}{\tau_p} + \frac{\beta \Gamma n(t)}{\tau_n}. \tag{3}$$

$$\frac{dn(t)}{dt} = \frac{I_p(t)}{q V_a} - G(t)\left[n(t) - n_0\right]p(t) - \frac{n(t)}{\tau_n}. \tag{4}$$

$$\frac{d\phi(t)}{dt} = \frac{\alpha}{2}\left[\Gamma v_g a_0 \left[n(t) - n_0\right] - \frac{1}{\tau_p}\right]. \tag{5}$$

In these equations Γ is the mode confinement factor, n_0 is the carrier density at transparency, τ_p is the photon lifetime, β is the spontaneous emission coupled into the lasing mode, τ_n is the carrier recombination time, q the electron charge, V_a is the active layer volume, v_g is the group velocity, a_0 is the gain coefficient, and α is the linewidth enhancement factor. The saturable gain G(t) can be expressed as:

$$G(t) = \frac{v_g a_0}{1 + \varepsilon p(t)}. \tag{6}$$

being ε the gain compression factor.

The input current $I_p(t)$ can be expressed as:

$$I_p(t) = I_{bias} + \sum_{n=1}^{N} m \cdot (I_{bias} - I_{th}) \cdot \cos(2\pi f_n t + \theta_n). \tag{7}$$

where m is the optical modulation index of the single channel, I_{bias} and I_{th} are the bias and threshold currents, respectively, f_n and θ_n are the nth carrier frequency and random phase, respectively.

The current $I_p(t)$ represents the electrical signal consisting of N frequency division multiplexed channels, and is generated by reverse fast Fourier transform (FFT) of N randomly phased tones. These signals represent generic broadband channels whose bandwidth varies between 20 and 150 MHz. We also consider a number of channels varying from 10 to 30.
The optical spectrum can be derived through FFT of the output intensity, which is directly derived by the photon density. Moreover, since the distortion is strongly dependent on the particular phases of the tones, we average the results of (20) sets of random phases.
Finally the spectrum of the signal, after passing a single mode standard fiber, is easily obtained by using the fiber transfer function reported in [3].
The optical signal transmitted from base station is transmitted to control station. In control station the optical signal is converted into RF signals by PIN-PD with sensitivity η and demodulate. When the number of input signal to LD is plural, intermodulation distortion (IMD), occurs by the combinations of input signals. Therefore, in control station, the received carrier to noise ratio is given as follows

$$CNR = \frac{\frac{1}{2} m^2 I_{ph}^2}{\left(RIN \cdot I_{ph}^2 + 2 \cdot e \cdot I_{ph} + \left\langle Ith^2 \right\rangle\right) \cdot B + NLD}. \quad (8)$$

where the noise terms in the denominator represent the relative intensity noise contribution from the laser, the shot noise term, the thermal noise from the receiver, and the equivalent noise accounting for non-linear distortion as a whole, respectively. In particular:

B : bandwidth per carrier
e : electron charge
RIN : the relative intensity noise for the laser diode
I_{ph} : average current received in control station
<Ith> : equivalent input noise current density in optical receiver
NLD : non-linear distortion power
m : optical modulation index

The term NLD accounts for all system contribution to non-linear distortions, such as laser clipping, chirping-fiber dispersion, and distortions due to the laser resonance frequency. While all the other noise contributions are calculated separately by their analytical expressions, NLD is calculated by simulation. In fact, the simulation allows the total spectrum at the receiving end to be evaluated. Consequently, the noise falling in the RF bandwidth of a given channel is integrated numerically to provide the valued of NLD. The worst case value of NLD is calculated and put in (8).

4 RESULTS

Figures 2 - 4 show the transmission performance of the system in terms of the CNR versus the optical modulation index (OMI) of the single channel.

It is worth noting that the increasing of the signal power, related with the OMI, produce an increasing of the CNR, until non-linear distortions become significant. This explains why there is an optimum value of the OMI allowing the highest CNR to be reached.

The impact of the number of RF carriers on the performance is depicted in figure 2, where the performance are reported for different values of the number of carriers, assuming a double-side bandwidth of 150 MHz for any single channel. In fact, it can be noticed that the CNR decreases with the increasing of the number of carriers. Basically, two reasons can be individuated for this: 1) the overall bandwidth of the RF signal increases and, consequently, the noise contributions grow; 2) the probability of clipping events (e.g. the clipping induced total distortion) increases when there are more carriers [4].

From these results it is possible to derive a first conclusion concerning the system design. Since the required CNR values, to obtain an error probability of 10^{-6}, using QPSK or 16 QAM modulation schemes, are 13.8 and 18.8 dB, respectively, we could say that it is possible to transmit up to 30 carriers using such modulation schemes. On the other hand, if the 64 QAM modulation is desired, the required CNR of 34.8 dB pose restrictions on the number of carriers and requires a limited range of value of the OMI (with 20 carriers the OMI should be in the range -26 ÷ -13 dB).

To investigate the impact of the single channel bandwidth on the system performance, figure 3 shows the CNR versus the OMI for three different values of the channel bandwidth, for 20 RF carriers. It is evident as the increasing of the bandwidth leads to an increasing of the noise contributions.

Taking into account the required values of CNR for the QPSK and 16QAM modulation schemes reported before, we could say that at least 20 broadband signals can be favorably transmitted in this kind of system.

In order to evaluate the effect of source chirping (in commingling with fiber dispersion) and of the resonance frequency of the laser on the system performance, it is useful to observe

figure 4. Here we assume to transmit 20 carriers. It is also presumed a fiber optic link length of 30 km. The RF signals comb, received from the antenna, is down shifted near the baseband (200 MHz) in one case, and to 1 GHz in the other case. In both the cases the performance are evaluated either considering the chirping effect and not.
By comparing the curves relating to the comb initial frequencies of 200 MHz and 1 GHz, respectively, we can observe a difference of few dB (3 in the worst case). This is due to the fact that the resonance frequency of the laser introduces further distortions when the laser operates to higher frequencies (the comb starts at higher frequency).
It is also to be highlighted that the impact of chirping effect is absolutely not negligible, since it brings about a penalty on the CNR of more than 13 dB. Again, considering the required values of CNR for QPSK and 16 QAM modulation schemes, we could say that, in this case, just a limited range of values for the OMI is admissible, considering a transmission of 20 carriers.

In order to provide some information about the feasibility of the system as a whole, it is worth while giving some information on the performance of the radio link too.
Assuming a transmitting power of 18 dBm, table 1 reports the outage probabilities for two different values of the cell area radius, and for the two considered modulation schemes, considering the meteorological statistics of the city of Roma, Italy [5].

5 CONCLUSIONS

An analysis of a fiber-to-the-air system has been carried out. The system performance is mainly affected by rainfall attenuation, concerning the radio link, and by noise contribution and non-linear distortions, regarding the optical link. In particular, optical non-linear distortions are due to non-linear characteristics of the laser source and to the intermingling of laser chirping and fiber dispersion.
The analysis revealed that this kind of system allows the transmission of some tens of broadband carriers, using two relevant modulation schemes: QPSK and 16 QAM.
In general, we could say that a tradeoff exists between the number of carriers and the channel bandwidth. Furthermore, the impact of chirping could be significant. The design of such a type of system must take into account all these effects. The model presented here allow the different effects to be taken into account.

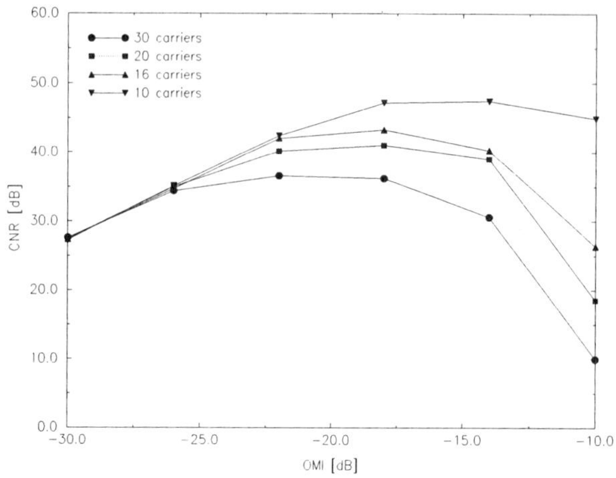

Figure 2. Carrier-to-noise ratio versus the optical modulation index, for different numbers of carriers. The RF comb starts from 200 Mhz, and the single channel bandwidth is 150 MHz.

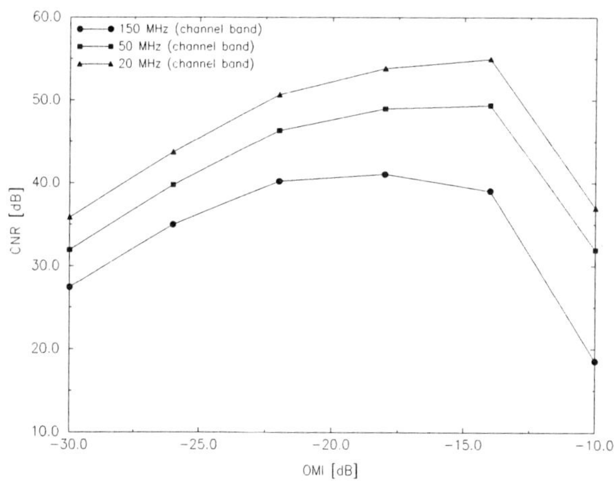

Figure 3. Carrier-to-noise ratio versus the optical modulation index, for different values of the channel bandwidth. The RF comb starts from 200 Mhz, and the number of carriers is 20.

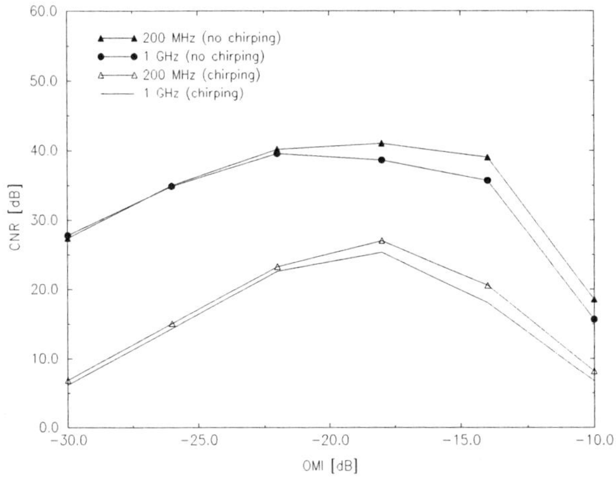

Figure 4. Carrier-to-noise ratio versus the optical modulation index, for different values of the start frequency and showing the chirping effect. The number of carriers is 20 and the channel bandwidth is 150 MHz.

OUTAGE PROBABILITY

Modulation	Cell radius: 300 m	Cell radius: 1 km
QPSK Modulation	10^{-6}	10^{-3}
16 QAM Modulation	10^{-5}	10^{-2}

Table 1. Outage probability as a function of the cell radius and of the adopted modulation scheme.

REFERENCES

[1] H. Harada, H-J Lee, S. Komaki, and N. Morinaga, "Performance Analysis of fiber-optic millimeter-wave band radio subscriber loop", IEICE Trans. Electron. Vol. E76-B, N^{O} 9, pp. 1128-1135, 1993.

[2] Luis M. Crreia, Ramjee Prasad, "An Overview of Wireless Broadband Communications", IEEE Comm. Mag. pp. 28-33 January 1997.

[3] C.-S. Li, F. F.-K. Tong, K. Liu, D. G. Messerschmitt, “Channel Capacity Optimization of Chirp-Limited Dense WDM/WDMA System Using OOK/FSK Modulation and Optical Filters” J. of L. Technology, vol.10, no8, august 1992.

[4] R. Sabella, M. Paciotti, A. Di Fonzo, "Analysis of clipping and chirping effects on AM-VSB CATV subcarrier multiplexed optical systems", in press on Journal of Optical Communications.

[5] F. Barbaliscia, A. Di Panfilo, Fondazione Ugo Bordoni, Internal Report no. FUB:1A07393, Roma, December 1993.

PART FOUR

Transmission in optical transport networks

Impact of Transmission Performances on Path Routing in All-Optical Transport Networks

M. Berdusco[+], S. Binetti[#], E. Iannone[+], M. Listanti, R. Sabella[#]*

[+] *Fondazione Ugo Bordoni, Rome, Italy*

* *INFOCOM Dpt., Università di Roma "La Sapienza"*

[#] *Ericsson Telecomunicazioni, R & D Division, Rome, Italy*

Roberto Sabella

e-mail: sabella@RD.tei.ericsson.se

tel: 06-72583046 fax: 06-72583940

Abstract

The impact of transmission related issues on the routing strategies for transparent all-optical WDM transport networks is analyzed in this paper. Three different categories of routing algorithms are analyzed: algorithms based on the Wavelength Path (WP) strategy, based on the Virtual Wavelength Path (VWP) strategy and requiring only a limited number of wavelength converters in the network (PVWP: Partial Virtual Wavelength Path). It results that the PVWP allows a saving in network devices with respect to the WP similar those permitted by the VWP also attaining transmission performances near those attained by the WP that are quite better that those attained by the VWP.

Keywords

WDM NETWORKS, TRANSPARENT NETWORKS, OPTICAL PATH ROUTING, WAVELENGTH CONVERSION, TRANSMISSION IN OPTICAL NETWORKS

1 INTRODUCTION

Optical transparent WDM networks seems the most promising candidates to constitute the future high-capacity-long distance communication infrastructure. In such networks the routing function is carried out directly in the optical domain by Optical Cross-Connects (OXC) so that high speed optical signals travels through the network without any optical-to-electrical conversion from their source to their destination. On this ground the concept of Optical Path (OP) can be defined, similarly to the concept of electrical path in SDH networks: an OP is a semi-permanent connection between path end points that is set-up utilizing OXCs [1,2].

Three different strategies have been proposed so far for the routing of optical paths in an optical WDM network: the Wavelength Path (WP) [1], the Virtual Wavelength Path (VWP) [1] and the Partial Virtual Wavelength Path (PVWP) [3,4]. In the WP, a fixed wavelength is assigned to each optical path and it is never changed in all its route through the network; in the VWP the wavelengths are assigned link-by-link to each path, so that the path wavelength can be changed at each OXC crossing; in the PVWP only a few OXCs in the network are able to change the wavelength of the crossing paths so that the path wavelength generally remain unchanged at the OXC crossing but for a limited number of changes at specific locations.

These classes of algorithms have been compared both considering the number of required WDM channels per fiber in the case in which a single fiber link exists between each couple of OXCs [5] and considering the required OXC dimension (number of i/o ports) in the case in which the maximum number of wavelengths per fiber is fixed [6,4]. The most significant comparison is the latter, since the number of wavelengths is generally constrained by the adopted technology; its main result is that the VWP and the PVWP allow a sensible save in OXC ports with respect to the WP. Moreover the PVWP reaches this task by using a limited number of wavelength converter devices, that is generally quite smaller that those required by the VWP.

However, transmission issues have never been considered while analyzing routing algorithms. While in an SDH network the transmitted signal is regenerated at each node, no regeneration is present in a transparent WDM network. Therefore noise and distortion accumulates along all the signal route. Since thousands kilometers long paths can be well present in a WDM network with a large geographical extension, perhaps not all the possible routes between a couple of path-end OXCs can be traveled with acceptable transmission performances. Thus, one more constraint is needed in the routing algorithms: all the routed paths must have acceptable transmission performances.

A comparison between WP, VWP and PVWP is for the first time presented in this paper. Specifically, a particular geographical network and a particular OXC technology are considered and different traffic volumes are routed through the network by using WP, VWP and PVWP algorithms. After the routing, each routed path is analyzed under the transmission point of view and the Unacceptable Paths Number (UPN: those path that do not assure acceptable transmission performances) is observed.

2 ROUTING ALGORITHMS

The routing algorithms used for the WP and the VWP scheme are those reported in [6]. These are based on an heuristic procedure of path accommodation and allows the OXC scale needed to accommodate a fixed traffic volume to be determined. As far as the PVWP is concerned, the so called global arrangement algorithm is used, that is described in [4].

All the routing algorithms accept as input a traffic pattern to accommodate; a traffic pattern is a set of path to be routed and it is represented as a matrix indicating how many paths have to be set-up between each pair of network nodes. A traffic volume, defined as the number of accommodated paths, is associated to each traffic pattern. In this paper we report average results with respect to the traffic pattern; that is we randomly generate a great number of traffic patterns that corresponds to the same traffic volume and average the considered figure of merit with respect to the traffic pattern.

Different figures of merit can be defined to compare routing algorithms. The first figure of merit is the average number of OXC ports N, that gives a measure of the needed OXC scale. Another important figure of merit is related to the number of wavelength converters used in the network: this is the "Wavelength Conversion Percentage Amount" (WCPA), defined as:

$$WCPA=\overline{K}/(N_1\ M), \quad (1)$$

where $\overline{K}$ is the average number of converters in the OXCs and M is the number of WDM channels per fiber. The figure of merit we mainly consider in this paper is the average number of paths that do not have acceptable transmission performances (UPN: Unacceptable Paths Number).

3 OXC ARCHITECTURE AND RELATED TECHNOLOGIES

To be able to consider transmission issues, the OXC architecture and the technology used to realize the adopted optical devices have to be specified.

In this paper we consider a consolidated OXC architecture, that can be also equipped with a limited number of wavelength converters without spoiling its functionality. This architecture, whose functional scheme is reported in figure 1, is based on the use of space switching optical matrixes, tunable filters and wavelength converters. In particular the routing is carried out by the space switching matrixes and the wavelength converters prevent conflicts between different paths entering the OXC at the same wavelength and directed towards the same OXC output port. Two Erbium-Doped Fiber Amplifiers are present to compensate the internal loss and part of the input link loss if it is the case.

The space switching matrixes are assumed to be realized with the thermo-optical technology [14], with a linear loss of 0.8/N (being N the number of input ports) and a crosstalk factor equal to χ=4 10^{-6}.

The tunable filters are assumed to be double-stage Fabry-Perot filter with a 3-dB bandwidth equal to 100 GHz and a loss of 2 dB.

The EDFAs are assumed to have a maximum output power of 18 dBm, a flat bandwidth of 35 nm and a noise factor F_E of 4 dB.

Finally the wavelength converters are assume to be based on Four-Wave Mixing in Semiconductor Optical Amplifiers (SOA) [9,10]. Since the behavior of the wavelength converters is quite important to determine the network transmission performances, it has to be described quite in detail. The model adopted to describe the performances of the wavelength converters is those reported in [9]. The wavelength converters parameters are relative to a low noise SOA, 1.5 mm long [15]. It is to be noted that, each time an optical signal experiences wavelength conversion, it is assumed that it is translated from lower limit of the WDM comb to the higher one, so to perform a worst case performance evaluation.

4 TRANSMISSION EVALUTATION

In order to evaluate the transmission performances of the optical path we assume that in-line optical amplification is used in the network, with an amplifier spacing L_A=50 km. The amplifier gain exactly compensates for the loss of the fiber link before the amplifier and the EDFAs inside the OXC compensate both the OXC internal losses and the loss of the fiber link between the last amplifier and the OXC. The fibers in the network are assumed to be Dispersion Compensating fibers with a dispersion D=4 ps/nm/km and an attenuation α=0.25 dB/km.

We assume that nonlinear fiber propagation can be neglected in our network. This means that the transmitted optical power has to be quite lower than the threshold power for the Brillouin and Raman effects and that the characteristic dispersion length L_D is always much smaller that the nonlinear length L_{NL} during signal propagation [16] so that even Kerr effect can be neglected.

Under these conditions, the main phenomena affecting the transmission performances of the optical paths are the ASE accumulation along the link and the in-band crosstalk inside the OXCs [15].

In this regime, the Q factor can be assumed as a good figure of merit to characterize the transmission performances of the optical paths. The Q factor is defined as the electrical signal-to-noise ratio after the baseband filter at the receiver; in the Gaussian approximation (that holds fairly well in our case) an error probability of 10^{-9} corresponds to an Q factor equal to 6. To assume a reasonable system margin, we will classify as unacceptable under the transmission point of view all the optical path for which the Q factor is lower than 6.5.

In our case, the Q factor has the following expression

$$Q = \frac{SNR_o}{\sqrt{2\,SNR_o\,b + \sqrt{b}}} . \qquad (2)$$

where SNRo is the optical signal to noise ratio in the signal bandwidth and b the ratio between the overall optical bandwidth and the electrical bandwidth (determined by the baseband filter after the receiver). We assume that the single channel bit-rate is equal to 2.5 Gbit/s and that b=5, condition that can be realized using an optical pre-detection filter with a bandwidth of 25 GHz. The channel spacing Δv is set to 300 GHz.

The Optical signal to noise ratio can be written as

$$SNR_o = \frac{\rho P_r}{2 R S_{tot}} . \qquad (3)$$

where ρ is the penalty due to the in-band crosstalk, P_r and S_{tot} are the peak optical power and the overall ASE power spectral density impinging the receiver.

The received optical power is simply proportional to the power at the output of each OXC and each EDFA by a factor depending on the loss of the path from the last OXC input to the receiver.

The penalty ρ can be evaluated by using the equivalent Gaussian noise approximation, as shown in [17]. In particular, ρ is a function of the overall crosstalk-to-signal ratio (CSR) whose expression in the worst case (wavelength translated signal against an unchanged crosstalk component) is simply given by

$$CSR = \sum_{k=1}^{NN} \left[\frac{|H(\Delta v)|^2}{\eta_k} + \chi M \right] . \qquad (4)$$

where NN is the number of traveled OXCs, H(v) is the frequency response of the tunable filters inside the OXC, Δv is the channel spacing, χ is the crosstalk factor of the space switching matrixes and M is the number of WDM channels per fiber. The factor η_k is equal to the wavelength converter efficiency if the considered path experiences wavelength conversion inside the k-th OXC, otherwise it is equal to one.

Finally the overall ASE power spectral density can be evaluated starting from the physical parameters of the different optical devices as detailed in [18]. The term S_{tot} depends heavily on the way in which the overall gain inside the OXC is divided between the two available EDFAs. In particular, minimizing S_{tot}, it can be demonstrated that the lowest noise power is obtained when the gain of the first amplifier is maximized. Thus, the amplifier at the OXC input has the maximum gain, that is determined by two constraints: first of all the maximum optical power the amplifier can emit in the signal bandwidth is limited (18 dBm in our case), second, the signal cannot be amplified so much that a possible output wavelength converter exists from a pump-probe working regime. This requires that, if P_p is the wavelength converter pump power, the signal power at the wavelength converter input is at least 10 dB smaller than P_p, thus limiting the gain of the input EDFA.

5 SIMULATION RESULTS

In order to provide simulation results, we consider in this paper a slightly modified version of the high speed Italian transport network. The network topology is shown in figure 2.

No optimization of the signal power and of the pump power of the wavelength converters is attempted in this paper. The signal power at the output of each OXC is fixed to 2 dBm, a sufficiently low value to neglect nonlinear effects during fiber propagation. The pump power of the wavelength converters is fixed to 8 dBm. Since this is the power injected into the SOA, this generally requires a pump laser emitting at least 14 dBm.

The average number of OXC ports versus the traffic volume is shown in figure 3 for the WP and the VWP algorithms. The results relative to the PVWP algorithm coincides with those relative to the VWP. The average number of ports increases at the increasing of the traffic volume and decreases at the increasing of the number of channels per fiber. Moreover, the little gain due to the adoption of wavelength conversion is evident from figure 3.

The average Wavelength Conversion Percentage Amount (WCPA) needed to implement the PVWP algorithm is shown in figure 4 versus the traffic volume. From the figure it results that the number of wavelength converters really needed to attain the performances of the VWP is quite smaller than 100%. Moreover the WCPA decreases at the increasing of the traffic volume. This is due to the fact that, increasing the traffic volume, the average number of OXC ports increases so that the routing map given by the WP is more and more similar that given by the VWP. This means that the number of needed wavelength conversions tends to decrease, as shown in figure 4.

The average Unacceptable Path Number (UPN) is shown in figure 5 versus the Normalized Traffic Volume (NTV), the traffic volume divided by the number of nodes, for the VWP algorithm and for M=4, 6. In the case M=8, UPN=100% for all the considered traffic volumes. Before discussing these results, an observation is needed. When we evaluate the transmission performances of an optical path including a certain number of wavelength converters, we assume that wavelength conversion really occurs at every converters with the maximum conversion range, i.e. with the minimum efficiency. This is a worst case approach that is far from reality for a large number of paths in the case of the VWP. If this approximation is removed and optimization of the transmitted power and of the pump power of the wavelength converters is carried out, better transmission performances can be obtained [19,20].

From figure 5, it can be observed that UPN increases at the increasing of the NTV. This is due to the fact that the average number of ports enlarges, thus increasing the OXC internal loss. Moreover, UPN rapidly growths with M, despite the decreasing of the average number of ports. This is due to two factors: the overall internal losses of the OXC increases and the efficiency of the wavelength converters decreases due to the greater conversion range. As far as the internal loss is concerned, it is the product of the losses of the input splitter, the filter, the SSM and the output combiner. The filter loss does not depend on M, the product of the other losses, assuming that the overall OXC capacity C=M N remains constant, depends on 1/M. Concerning the efficiency, in our case, for M=4 it is about -6.8 dB, for M=6 it is -11.3 dB and for M=8 it is -14.7 dB.

The dependence of UPN on NTV in the case of the PVWP algorithm is shown in figure 6. The curves for M=4 is not reported in the figure since UPN=0 in all the considered cases. Since the number of wavelength converters in the network is by far smaller than in the case of the VWP, even UPN is quite smaller. With M=6 UPN is smaller than 2% up to NTV=150.

6 CONCLUSIONS

Transmission problems cannot be neglected in designing routing algorithms for wide area transparent WDM networks. In this paper, the impact of the main transmission issues on the path routing is analyzed in the case of a typical wide area network (a modified form of the high capacity Italian network) assuming to route optical paths by Optical Cross-Connects based on space switching and allowing, if the case, even wavelength conversion. Wavelength converters based on Four-Wave Mixing in Semiconductor amplifiers are considered. While in the case of routing based on the Wavelength Path scheme, the transmission issues are not important, they have to be considered in the case of the adoption of the Virtual Wavelength Path or of the Partial Virtual Wavelength Path schemes. In this cases, it can occur that some optical path do not have acceptable transmission performances. Two ways are possible to cope with this problem: the unacceptable paths can be routed again on a different route presenting good transmission performances by further enlarging the OXC scale or electronic regenerators can be put on the unacceptable path, so spoiling the network optical transparency.

Comparing the VWP and the PVWP, it results that the number of unacceptable paths is quite larger in the case of the VWP, due to the larger number of wavelength converters crossed by each path. This is another important advantage of the PVWP, besides the possibility to attain the same routing performances of the VWP with simpler OXCs.

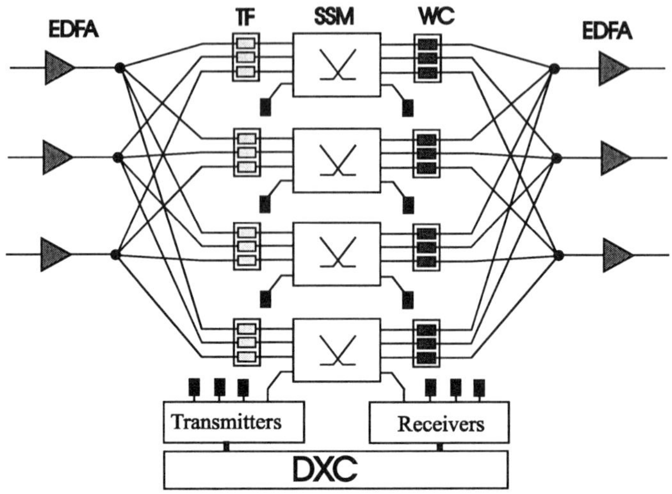

Figure 1 OXC Architecture

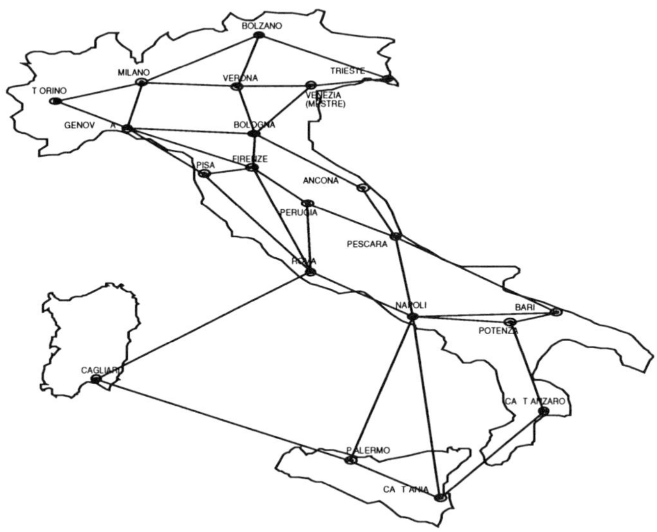

Figure 2 Topology of the high-capacity Italian network used in this paper.

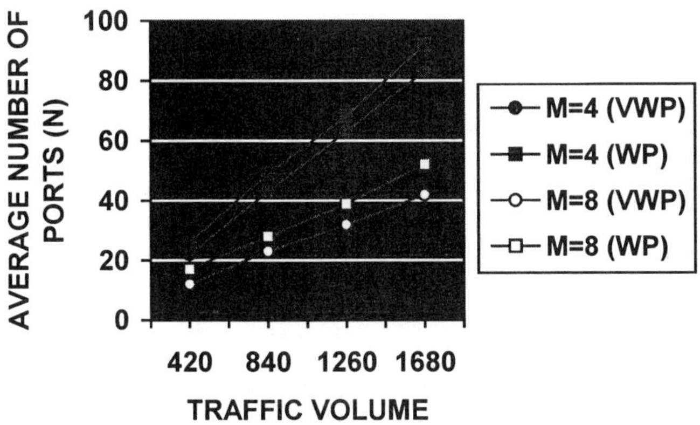

Figure 3 Average number of ports versus the traffic volume for the VWP algorithms.

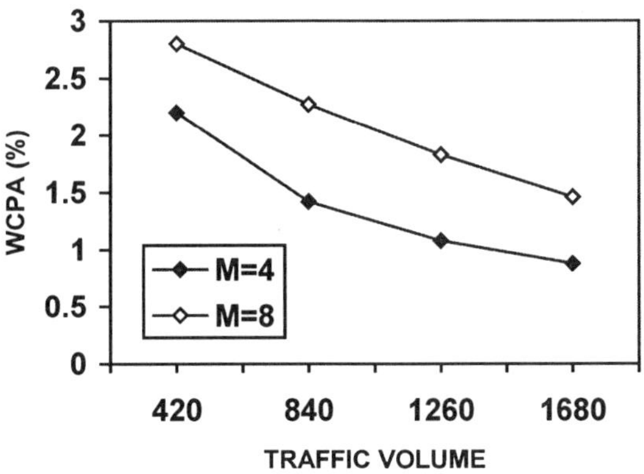

Figure 4:Average Wavelength Conversion Percentage Amount (WCPA) versus the traffic volume.

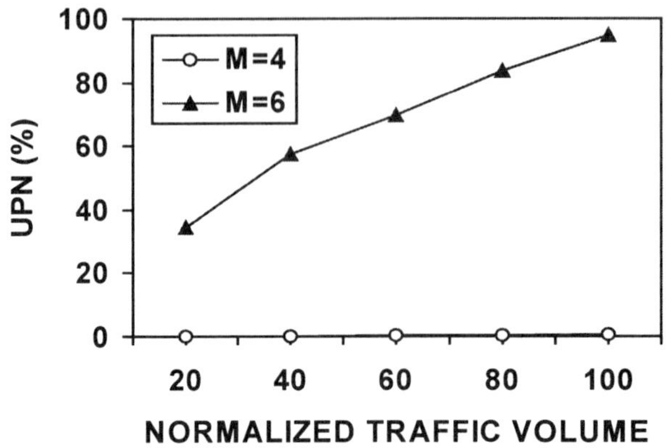

Figure 5:Average Unacceptable Path Number (UPN) versus the Normalized Traffic Volume for the VWP algorithm.

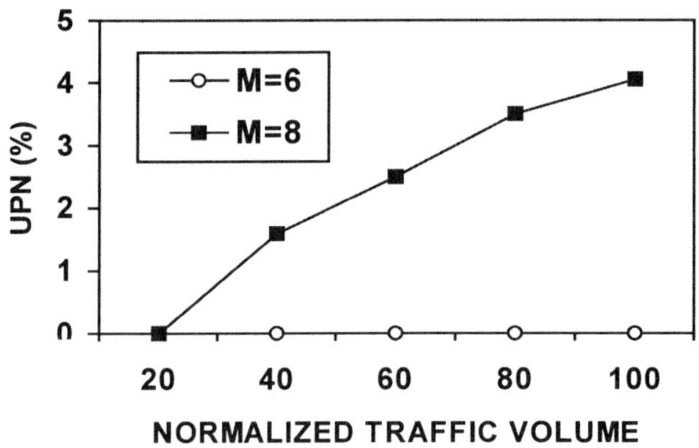

Figure 6:Average Unacceptable Path Number (UPN) versus the Normalized Traffic Volume for the PVWP algorithm.

REFERENCES

[1] K. Sato, S. Okamoto and H.Hadama, "Network performance and integrity enhancement with optical path layer technologies", IEEE-Journal on Selected Areas in Communications, vol. 12, no. 1, pp. 159-170, 1994.

[2] R. Ramaswami, K.N. Sivarajan, "Routing and Wavelength Assignment in All-Optical Networks", IEEE/ACM Transaction on Networking, vol. 3, no. 5, pp. 489-500, 1995.

[3] M. Listanti, M. Berdusco, R. Sabella, "Optical Path Strategies in WDM All-Optical Networks: Minimization of Wavelength Converters in Optical Cross-Connect", GLOBECOM '97, November 1997, Phoenix, AZ.

[4] M. Listanti, M. Berdusco, R. Sabella, "A New Strategy For Employing Wavelength Conversion in WDM Optical Networks", LEOS '97, November 1997, San Francisco.

[5] N. Nagatsu, Y. Hamazumi, K. Sato, "Optical Path Accommodation Designs Applicable to Large Scale Networks", IEICE Transactions on Communications, vol. E78-B, n° 4, pp. 597-607, 1995.

[6] N. Nagatsu, et al., "Optical Path Cross-Connect System Scale Evaluation Using Path Accommodation Design For Restricted Wavelength Multiplexing", IEEE Journal of Selected Areas on Communications, vol. 14, no. 5, pp.893-902, 1996.

[7] A. Watanabe, S. Okamoto, K. Sato, "Optical path cross-connect node architecture with high modularity for photonic transport networks", IEICE Transaction on Communications, vol. E77-B, no. 10, pp. 1220-1229, 1994.

[8] G.R. Hill et al., "A transport network layer based on optical network elements", IEEE-Journal of Lightwave Technol., vol.11, no.5/6, pp. 667-679, 1993.

[9] A. D'Ottavi, E. Iannone, A. Mecozzi, S. Scotti, P. Spano, R. Dall'Ara, J. Eckner, G. Guekos,"Efficiency and Noise Performances of Wavelength Converters Based on FWM in Semiconductor Optical Amplifiers", IEEE-Photonics Technology Letters, vol. 31, no. 4, 1995.

[10] R. Ludwig, G. Raybon, "BER Measurements of Frequency Converted Signals using Four-Wave Mixing in a Semiconductor Laser Amplifier at 1, 2.5, 5 and 10 Gbit/s"; Electronics Letters, vol. 30, n° 4, pp. 338-339, 1994.

[11] S. J. B. Yoo, C. Caneau, R. Bhat, M. A. Koza, A. Rayhil, N. Antoniades, “All-Optical Wavelength Conversion by Quasi-Phasematched Difference Frequency Generation in AlGaAs Waveguides”; Applyed Physics Letters, vol. 68, pp. 2609-2610, 1966.

[12] M. Bischoff, M. N. Huber, O. Jahreis, F. Derr, F. Hulm, “Operation and Maintainance for an All-Optical Transport Network”; IEEE-Communication Magazine, vol. 11, pp. 136- 142, 1996.

[13] T. Durhuus, B. Mikkelsen, C. Joergensen, S. L. Danielsen, K. E. Stubkjaer, "All-Optical Wavelength Conversion by Semiconductor Optical Amplifiers",.IEEE-Journal of Lightwave Technology, vol. 14, n° 8, pp. 942-954, 1996.

[14] M. Koga, A. Watanabe, S. Okamoto, K. Sato, M. Okuno, "8x16 delivery-and-coupling-type optical switches for a 320-Gb/s throughput optical path cross-connect system", OFC '96 Technical Digest, Paper ThN3, 1996.

[15] E. Iannone, R. Sabella, L. de Stefano, F. Valeri, "All-optical wavelength conversion in optical multicarrier networks", IEEE Transactions on Communications, vol. 44, no. 6, pp. 716 - 724, 1996.

[16] G. P. Agrawal, *Nonlinear Fibre Optics,* Academic Press, New York, 1989.

[17] L. Moura, N. Karafolas, P. Lane, A. Hill, J. O’Reilly, “Modeling of Interferometric Crosstalk in Optical Networks”; Proceedings of GLOBECOM ’96, London, November 1996, pp. 333-337.

[18] E. Iannone, R. Sabella, "Performance Evaluation of an Optical Multi-Carrier Network using wavelength Converters based on FWM in Semiconductor Optical Amplifiers", IEEE-Journal of Lightwave Technology, vol. 13, no. 2, pp. 312-324, 1995.

[19] R. Sabella, E. Iannone, E. Pagano, "Optical Transport Networks Employing All-Optical Wavelength Conversion: Limits and Features", IEEE Journal of Selected Areas in Communications, vol. 14, no. 6, 1996.

[20] E. Iannone, R. Sabella, L. de Stefano, F. Valeri, "All-optical wavelength conversion in optical multicarrier networks", IEEE Transactions on Communications, vol. 44, no. 6, pp. 716 - 724, 1996.

Electrical Dispersion Compensation for 10 Gbit/s Transmission Systems : Simulation Results

Sven Otte, Werner Rosenkranz
University of Kiel
Kaiserstraße 2
24143 Kiel, Germany
Tel.: +49 431 77572-756, FAX -753
email : svo@techfak.uni-kiel.de

Abstract

In this paper we will compare two electrical signal processing techniques to mitigate the influence of chromatic dispersion in long-haul fiber optic systems at 10Gbit/s. The presented techniques are easy to integrate in existing systems and can significantly improve the dispersion limited distance in transmission spans. In particular we analyse the performance of the decision feedback (DF) equalizer with only one or two coefficients and we will compare the performance with a linear equalization filter. We describe methodes to determine the coefficients of the DF as well as an optimal filter design in the mean square error (MSE) sense for the equalization filter.

Keywords

Electrical dispersion compensation, Decision Feedback equalization

1. INTRODUCTION

Electrical signal processing at the receiver improves the performance without modifying the transmission format. Furthermore it can be adaptive to time varying sources or channel conditions. The optimal receiver for compensating intersymbol interference (ISI) is the Maximum-likelihood-detector [Winters, 91]. However, a realization with today's technology is not practical in high speed transmission systems. The decision-feedback detector is closely related to the Maximum-liklihood detector, but it is very simple to realize and can significantly reduce the effect of ISI due to chromatic and polarization dispersion in optical fibers as well as interference between signals in a wavelength division multiplex system, e.g. interference due to nonlinearity. In section 2 we will briefly describe the simulation model and will then give a short introduction to the theoretic backround of the DF as well as the linear equalization (section 3). In order to investigate the performance of the proposed equalization techniques we will only consider the chromatic dispersion in the fiber as the main source of distortion. In section 4 we discuss the improvement of the eye opening and the bit error ratio (BER).

2. SIMULATION MODEL

A filtered nonreturn to zero (NRZ) input data stream is used as a modulating signal which is assumed to be chirp-free. The complex envelope at the fiber input is $y_T(t)$ which is propotional to the square root of the optical laser output power. As mentioned the dominant source of distortion is the chromatic dispersion in the fiber. The frequency response of the fiber is given by $H(f) = exp(-jbf^2) \overset{\mathcal{F}}{\bullet\!\!-\!\!\circ} h(t)$, $b = \pi D(\lambda_T)\lambda_T^2/cL$, with $\lambda_T = 1550nm$, $D(\lambda_T) = 17ps/nm/km$ for standard single mode fibers (SSMF), L the fiber length and c the speed of light. The fiber output signal is $y_r(t) = h(t) * y_T(t)$. The PIN diode at the receiver converts the electrical field into an electrical signal proportional to the signal power $y(t) = |y_r(t)|^2$. The overall system is thus nonlinear due to the square root and the magnitude squared operation even though the fiber itselfs is a linear filter.

3. COMPENSATION TECHNIQUES

3.1 Decision feedback equalization

The decision feedback equalization is a nonlinear technique to compensate for both linear and nonlinear distortions. Already detected bits are used to generate an error signal $e(k)$ which ist substracted from the sampled imput signal $y(k)$. Figure 1. shows a block diagram of a 2nd order DF-detector. For better comprehension we will consider a linear transmission system with the overall impulse response $g(t)$. The noisefree receiver input baseband signal is : $y(t) = \sum_{l=-\infty}^{\infty} d_l \cdot g(t - l \cdot T_b)$, where $d_l \in \{0, 1\}$ are the transmitted bits and $f_b = 1/T_b$ is the symbol rate. Sampling this signal gives $y(t_0 + kT_b) = d_k g(t_0) + \sum_{\substack{l=-\infty \\ l\neq 0}}^{\infty} d_{k-l} \cdot g(t_0 + l \cdot T_b)$ where t_0 denotes the sampling instant of the impulse response $g(t)$. If $g(t) = 0$ for $t < 0$ the actual weighted bit to be decided is $d_k \cdot g(t_0) = y(k) - \sum_{l=1}^{\infty} d_{k-l} \cdot g(t_0 + l \cdot T_b)$ where the 2nd term is the error signal $e(k)$. Note that the eye-opening is given by $V_{DF} = g(t_0)$ and the decision threshold is $S = g(t_0)/2$. Knowledge of the impulse response and of the previously detected bits is sufficient to remove ISI completely. Even if the impulse response is well known, the

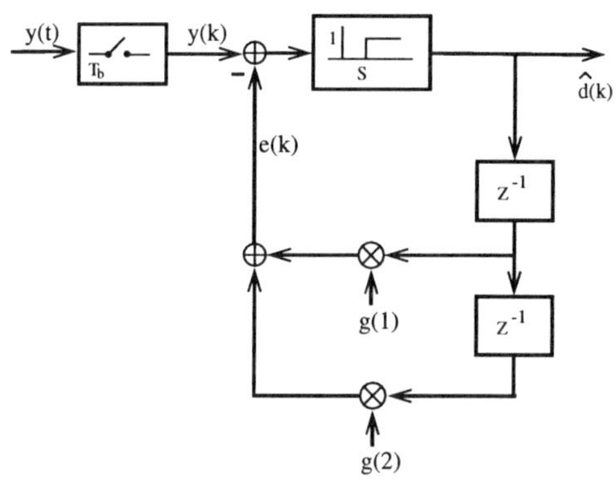

Figure 1: DF detector 2nd order

performance of the DF is a function of the sampling instant t_0.

3.2 Linear equalization

As an alternative to the DF a linear equalization filter can be used to compensate for the linear part of the distortion. The key issue for designing this filter is the knowledge about the impulse response of the system. We will first estimate the impulse response in the mean square sense by a given sequence of the received signal and the original data input stream and will then design the equalization filter.

Let $y(t)$ be the receiving signal (e.g. measurement results) and $\hat{y}(t)$ the output of the estimated linear system $g_e(t)$, where index e denotes estimate. For simulation these signals are sampled by the sampling rate $f_s = N \cdot f_b$, where N is the oversampling factor. We define the following vectors :

$$\mathbf{D} = \begin{bmatrix} d_0 & 0 & 0 & 0 & \cdots \\ 0 & d_0 & 0 & 0 & \cdots \\ \cdots & \cdots & \cdots & \cdots & \cdots \\ d_1 & 0 & d_0 & 0 & \cdots \\ 0 & d_1 & 0 & d_0 & \cdots \\ \cdots & \cdots & \cdots & \cdots & \cdots \end{bmatrix} \quad \begin{array}{l} \mathbf{d} = [d_0 d_1 \cdots d_{N_d}]^T \\ \mathbf{g_e} = [g_e(0)\; g_e(1) \cdots g_e(L \cdot N)]^T \\ \hat{\mathbf{y}} = \mathbf{D} \cdot \mathbf{g_e} \\ \mathbf{y} = [y(0)\; y(1) \cdots y((N_d + L)N)]^T \end{array}$$

$\mathbf{D}$ is the convolution matrix of the input data $d(k)$, $\mathbf{g_e}$ is the vector of the linear impulse response which is set to zero for $t > L \cdot T_b$ and N_d is the number of transmitted data. Minimizing $\boldsymbol{\epsilon}^T\boldsymbol{\epsilon} = (\mathbf{y} - \mathbf{D} \cdot \mathbf{g_e})^T (\mathbf{y} - \mathbf{D} \cdot \mathbf{g_e})$ which is the square sum of the estimating error gives

$$\mathbf{g_e} = \left(\mathbf{D}^T\mathbf{D}\right)^{-1} \mathbf{D}^T\mathbf{y} \tag{1}$$

With the estimated impulse response we can now design the equalization filter $e(t)$. An ideal transmission channel is $c(t) = \delta(t - t_d)$ where t_d is the signal time delay. To reach nearly ideal conditions we set $e(t) * g_e(t) = \delta(t - t_d) + \epsilon(t)$ where $\epsilon(t)$ is an error signal. The time continuous functions are sampled with the sampling rate f_s. Let $\mathbf{G}$ be the convolution matrix of the estimated linear system so that $\mathbf{G} \cdot \mathbf{d} = \hat{\mathbf{y}}$ and $\mathbf{e} = [e(0)\; e(1) \cdots e(N_e)]^T$ the vector of the equalization filter. The optimal equalization filter in the mean square sense is then given by

$$\mathbf{e} = \left(\mathbf{G}^T\mathbf{G}\right)^{-1} \mathbf{G}^T\boldsymbol{\delta} \tag{2}$$

which is the same as equation 1. Note that $N = 1$ gives the well known taped delay line [Winters 90].

4. RESULTS

Figure 2.a shows the normalized eye opening versus the fiber length. Without equalization the eye is totally closed after 140 km due to chromatic dispersion in the fiber and nonlinear properties of the PIN envelope demodulation.

Accepting an eye opening of 50 % in the considered case, the repeaterless distance could be increased from 90 km to more than 125 km for a 2nd order DF equalizer. In Figure 2.b simulation results of the BER applying the differ-

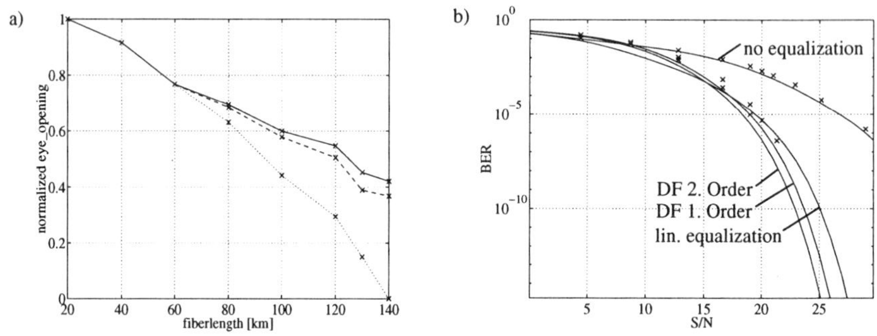

ent approaches are shown. The BER is determined by adding white gaussian noise and counting the occured errors (Monte Carlo Simulation). An alternative approach to determine the BER is the quasianalytic method [Jeruchim, 92]. In a practical range ($\sim 10^{-12}$) the DF-equalizer is more efficient then the linear equalizer filter. Note that most of the ISI is removed with only one feedback coefficient. For very low S/N the performance of the DF-equalizer decreases due to error propagation.

5. SUMMARY

The performance of a DF-equalizer of 1st and 2nd order is presented and compared to a linear equalizer in a simulated 10Gbit/s lightwave system. We evaluated the performance both in terms of eye opening and BER. Simulation show that equalization in the electrical part of the receiver can significantly compensate for distortion due to chromatic dispersion. It has been shown, that for the dispersive channel DF is superior to linear equalization. Moreover DF allows simpler realization. It is proposed, that future receiver chip designs should contain the very simple DF-structure.

6. REFERENCES

Winters, J.H. and Gitlin, R.D. (1990) Electrical signal processing techniques in long-haul fiber-optic systems. *IEEE Transactions on Communications, Vol. 38*

Winters, J.H. and Kasturia, S. (1991) Constrained maximum-likelihood Detection for high-speed fiber-optic systems. *Globecom 1991*

M. C. Jeruchim, P. Balaban, K. S. Shanmugan (1992) Simulation of Communication Systems. *Plenum Press New York and London*

PART FIVE

Transmission in optical transport networks II

Modelling of system impairments due to Four Wave Mixing in Dispersion Shifted links having non uniform zero-dispersion wavelength

Ernesto Ciaramella

CSELT- Centro Studi E Laboratori Telecomunicazioni

Via G.Reiss Romoli, 274 (10148) Torino ITALY

Phone: +39 11 2285442 Fax +39 11 2285840

e-mail: ernesto.ciaramella@cselt.it

Abstract.

Four Wave Mixing impairments in geographical WDM networks strongly depend on the zero dispersion wavelength of the fibre, that in actual links is far from being uniform. Here the role of non-uniform chromatic dispersion is investigated for the first time, using a new simplified and effective method. Obtained results show a decrease of the system penalty for increased fluctuations. However this is not a monotonic behavior, as resonance peaks can be observed. Monte Carlo simulations are also carried out, that give for the first time the estimation of the probability distribution function of the system penalty.

1 INTRODUCTION

Presently, it is widely accepted that optical networks would exploit Wavelength Division Multiplexing (WDM) technique. However, when implementing a WDM network, particular care should be taken in order to avoid the occurrence of a non linear propagation effect, namely the Four Wave Mixing (FWM). Indeed FWM could dramatically affect system performance, whenever the chromatic dispersion coefficient of the fibre is very low (Tkach, 1995); this is not the case of most of the regional and national fiber networks, that are typically made of conventional single mode fibre (G.652), having a non-negligible chromatic dispersion coefficient (D=17 ps/nm/km). Actually, FWM has to be considered only in Dispersion Shifted (DS) fibers (G.653), which are usually deployed in the long international links, and also, in some countries, in the national transport network. Even in those fibers, FWM limitations strongly depend on the chromatic dispersion, or the zero-dispersion wavelength (λ_0), of the fiber, so that, in principle, system impairments could be strongly reduced by a slight detuning from the optimum phase matching condition. In a uniform fiber this could be accomplished by setting the channels far enough from the λ_0 (Kurtze, 1994). However installed DS links are actually far from being uniform. Indeed they are made of different jointed spans, and the λ_0 value of those spans is randomly distributed according to the manufacturer statistical distribution. Thus the chromatic dispersion properties of those links can not be determined. Although FWM relevance can not be simply estimated in non-uniform fibers, FWM limitations over a given link are expected to be quite different from what would be observed in the corresponding uniform fiber case. Moreover, when considering a whole DS fiber network, FWM impairments should be rather investigated using a statistical approach, using Monte Carlo simulations. Unfortunately, a thorough analysis of the relevance of chromatic dispersion non-uniformity still lacks, mainly because it would have required a large amount of simulations, and those would have been much too time consuming if carried out by means of the usual numerical approach (Agrawal, 1995).

Here we present a new method for determining system penalties in WDM channels in non-uniform fiber links. This approach is used to evaluate FWM impairments in links with non-uniform chromatic dispersion. First, we considered a given link, and we adiabatically scaled the fluctuation magnitude: this alters FWM efficiency, but it does not monotonically affect the system penalties. Indeed, resonance peaks are found, arising from the partial phase matching produced in different spans. Occurrence of these peaks is observed even for relevant standard deviation of the λ_0. Monte Carlo simulations are also carried out, that give a quantitative estimation of the probability density function of the system penalty.

2 MODEL AND RESULTS

Due to the relevant computing time of the standard method (Agrawal, 1995), it was necessary to implement a new, simplified tool in order to fast evaluate the eye closure of the channels affected by FWM. To this aim we developed a simplified analytical approach. For the sake of simplicity, we did not consider in-line amplifiers, and assumed all parallel polarisations, that maximise the FWM impairments. We also concentrated on partially degenerate FWM, also known as Three Wave Mixing (Wu, 1995, Cappellini 1991). It should be noted that this is likely the first relevant parametric process in WDM systems; indeed, in actual systems either the polarisations and the frequencies of the carriers do exhibit random fluctuations, thus a three wave coupling is far more likely to occur than the completely non-degenerate FWM.

Under the previous assumptions, system impairments were calculated using this tool. The method is based on the analytic solution of the coupled mode equations (Cappellini, 1991), neglecting the (low) energy transfer from the three signals to other frequencies. Although this approach requires much shorter computing time (by a factor 10^3), tool results show very good agreement with those obtained by means of the most proper split-step simulations (Agrawal, 1995). This can be seen in Figure 1, reporting comparative results for the eye closure dependence on D in uniform links (continuous curve: tool; circles: standard split-step simulations).

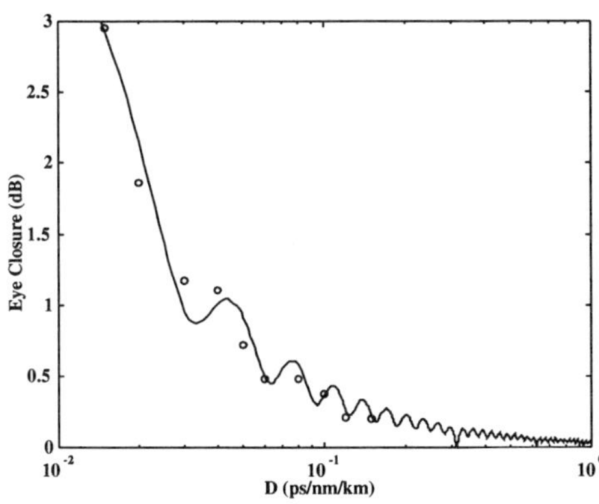

Figure 1 Comparison of the analytical tool results (line) and those obtained using a standard split-step method (circles). Results refer to the uniform case.

This tool was used to estimate the FWM penalty in non-uniform DS links for the first time. We considered links composed of 3 km long uniform spans and with 0.24 dB/km loss coefficient. Calculations were stopped to 30 km length, as after this distance propagation is almost linear, because of the fibre attenuation (in this case the non linear effective length would be less than 20 km). The λ_0's were randomly chosen according to a gaussian distribution with $<\lambda_0>$ and σ parameters (respectively, the average zero dispersion wavelength and the standard deviation). Thus the chromatic dispersion at the signal wavelength λ_S was calculated by the usual

$$D(\lambda_S) = S(\lambda_S - \lambda_0) \tag{1}$$

where S is the dispersion slope (S=0.07 ps/nm^2/km). The effect of the fluctuations was first modelled by considering a simplified case with adiabatic switching from the uniform fibre. A set of ten $\Delta\tilde{\lambda}$ was sorted, according to a normal Gaussian distribution. The λ_0's were calculated, by scaling according to:

$$\lambda_0 = \lambda_0 + \sigma(\Delta\tilde{\lambda}). \tag{2}$$

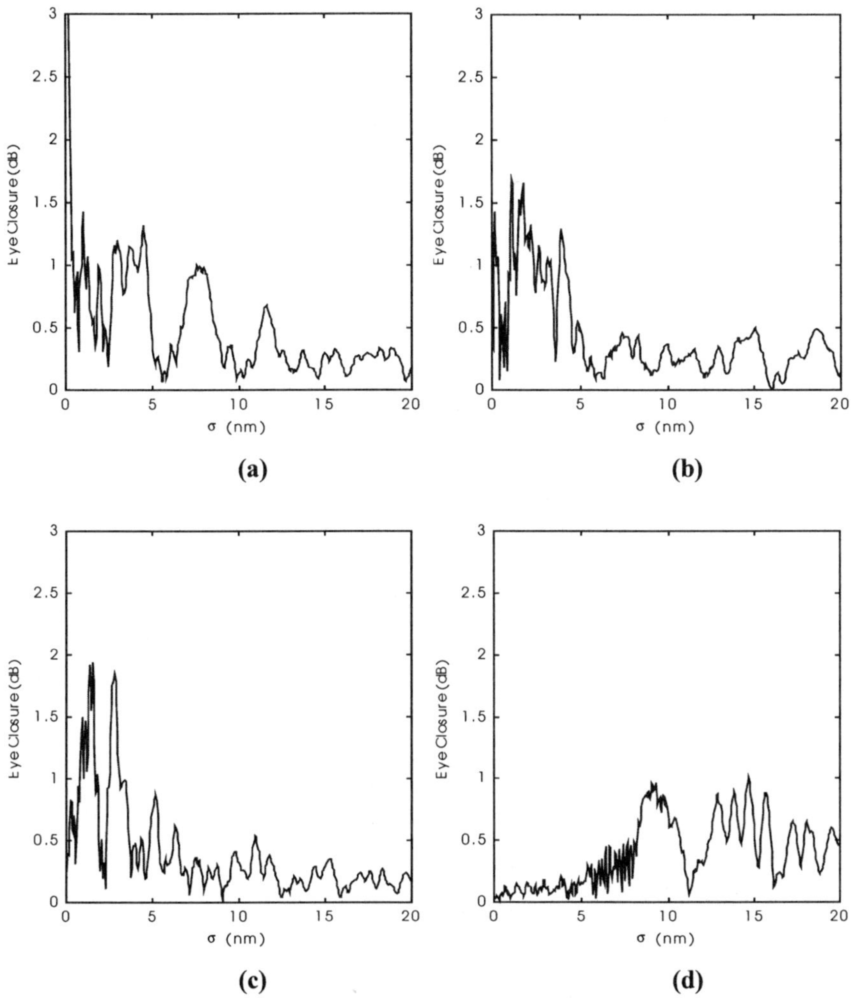

Figure 2 Effect of adiabatic increase of the zero dispersion wavelength fluctuations. The four figures report the eye closure dependence on σ. The average chromatic dispersion coefficients are respectively 0 (a), 0.1 (b), 0.2 (c) and 1 ps/nm/km (d). The WDM channels are equally spaced by 400 GHz.

According to Equation 2, the fluctuations were adiabatically switched on starting from σ=0 (uniform case) up to 20 nm (quite higher than observed values), with 0.1 nm step. For every σ value the eye penalty was evaluated for the three interacting channels of a WDM system with 400 GHz channel spacing and 4mW/channel optical power. Figure 3 shows the typical results obtained by scaling the σ value. In all the figures we report only the eye closure penalty referred to the central channel, which is typically the most affected. It is expected that fluctuations can mitigate the system impairments, with respect to the uniform case. This effect can be observed in Figure 2.a, that indeed refers to the average chromatic dispersion value <D>=0 ps/nm/km.

Here for σ=0 nm perfect phase matching occurs, giving a quite high penalty (more than 3 dB eye closure). When fluctuations increase phase matching is affected, and the penalty decreases. It is noteworthy, the effect over the system is clearly not monotonic; notable resonance peaks are indeed obtained, resulting from partial phase matching.

On the other hand, when we set a non-zero average dispersion value, different behaviours were obtained (see Figures 2.b and 2.c). Again, the fluctuation effect decreases the FWM impact, but here the phenomenon is less significant. Indeed still for large fluctuations (σ>10 nm) FWM can originate system penalties not much lower than in the corresponding uniform case (σ=0 nm). Moreover, we found a quite surprising result for a still higher <D> value (Figure 2.d). Here chromatic dispersion fluctuations make the FWM more effective and many penalty peaks can be observed for $\sigma \approx$10-15 nm. Clearly, the behaviours shown in Figure 2 are rather general, but details of the curves are actually dependent on the initial $\Delta\tilde{\lambda}$ set.

In order to obtain a more quantitative assessment of the fluctuation relevance with respect to the uniform worst case, Monte Carlo simulations were eventually performed assuming σ=5 nm (a realistic value) and <*D*>=0 ps/nm/km. System penalties were calculated over statistical ensembles of 200 DS fibre links, and an estimation of the probability distribution function was derived. Figure 3 shows the obtained histogram. The worst case penalty (more than 3 dB) has negligible penalty, but high penalty values (>1 dB) are still likely. On the other hand, there is also a significant probability that fluctuations can decrease the penalty to negligible values (<0.2 dB).

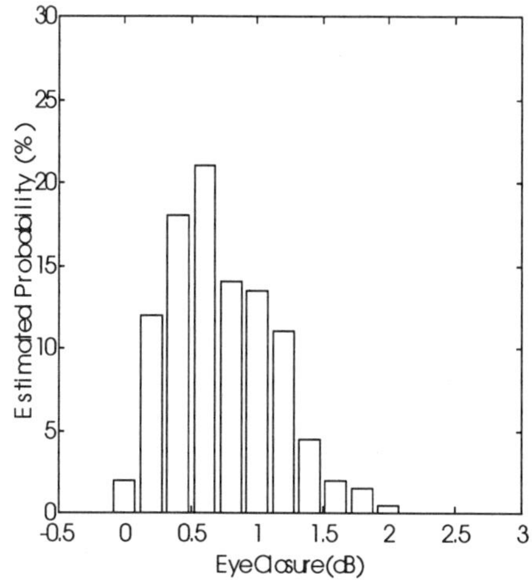

Figure 3 Probability histograms of the eye closure penalty for P=4 mW/channel. The average chromatic dispersion coefficient is <D>=0 ps/nm/km and the standard deviation is σ= 5 nm.

CONCLUSIONS

We investigated the role of FWM in WDM systems in non-uniform DS links, by means of a new, effective tool, considering three signal interaction. It was demonstrated for the first time how the chromatic dispersion fluctuations can modify FWM limitations with respect to the uniform fiber case. Fluctuations generally diminish the impairments with respect to the worst case (corresponding to perfect phase matching), but resonance peaks can be produced by partial phase matching. This effect is confirmed by Monte Carlo simulations, showing a quite broad probability distribution function of the FWM penalty in a realistic DS fiber network. Obtained results indicate a statistical reduction of the penalty. Although in some cases there is a quite complete suppression of the nonlinear effect, FWM can still produce relevant system impairments.

REFERENCES

Agrawal G. P, *Nonlinear fiber optics*, Academic Press, 1995.

Cappellini, G., Trillo S., "Third order three wave mixing in single mode fibres: exact solution and spatial instabilities", J. Opt. Soc. Am. B, 8, 824-838 (1991)

Fukui M., Aisawa S., Ishida O., Shimano K., Umeda A., Sakamoto T., Oda K., Takachio N. , "Allowable fibre input power for installed dispersion shifted fibre in equal/unequal channel spaced WDM transmission systems", *El. Lett.*. 33, 639-640 (1997).

Kurtze C., "Suppression of fiber nonlinearities by appropriate dispersion management", *IEEE Photonic Technol. Lett.*, 5, 1250-1253 (1993).

Tkach R., Chraplyvy A.R., Forghieri F., Gnauck A.H., Derosier R.M., "Four Photon Mixing and high speed WDM systems", *J. Lightwave Technol.*, 13, 841-849 (1995)

Wu W., Yeh P., "Energy coupling by partially degenerate Four Wave Mixing in multichannel lightwave systems", Photon. Technol. Lett., 7, 545 (1995).

Investigation on the capacities of optically amplified transmission systems in Geographic Networks

F. Matera, M. Romagnoli, M. Settembre, L. Socci
Fondazione Ugo Bordoni, via B. Castiglione 59, 00142 Roma ITALY
tel. +39-6-54802215, fax +39-6-54804402, e-mail: mat@fub.it

Abstract. In this work we present a survey of the main optically amplified communication systems that have been proposed to achieve a very high capacity transmission and their performances are analysed by means of the numerical simulations. Assuming different link lengths, guidelines for the implementations of high capacity systems are shown.

1. INTRODUCTION. Telecommunication field has shown a deep evolution in the last years, mainly due to the invention of the optical fibres [1] and other devices as lasers and optical amplifiers [1]. Such an evolution has been demonstrated by means of several experiments that have shown how to transmit signals with a capacity of hundreds of Gbit/s on transoceanic distances [2-4].

At the moment several kinds of optical systems have shown their potentiality, but it is very difficult to identify the best optical system, since such a choice depends on several factors as the physic parameters of the link, the cost, the required information traffic and the transparency of the information.

In this work we present a survey, based on the results found in the environment of the European COST239 project and in the ACTS ESTHER and UPGRADE projects, on the potentiality of the main optical communication systems that have been proposed in the environment of the high capacity transmission. The results have been obtained by means of numerical simulations and the main characteristics of the simulation tools can be found elsewhere [5-6].

Numerical simulation has been a fundamental tool for the study of the optical communication since the analytical theories are not able to take into consideration all the effects that are simultaneously present in a real link. The numerical code used in this work permits to evaluate the performance of optically amplified digital systems including all the effects that are present in a real systems. Such a code can be used also for the study of optical transport networks when devices as splitters, Add/Drop Multiplexing (ADM), Wavelength Converters (WC) and Optical Cross Connect (OXC) are considered. Such a code has been previously checked with other groups of the ACTS projects and it has been found an excellent agreement by the comparison with analytical theories and a good agreement with experiments. This has permitted to consider the simulation tools very reliable.

The results regards the propagation at 1.5 μm considering both DS and step-index fibers using Erbium Doped Fiber Amplifiers (EDFA). For step-index fibers is also investigated the possibility to operate at 1.3 μm by using Semiconductor Optical Amplifiers (SOAs) [6].

2. MODEL OF THE OPTICAL SYSTEMS. An optical communication is obtained by means of a *transmitter*, a *transmission line* and a *receiver*. According to the modern concept on the optical communication the transmission line is supposed to be composed by optical fibres and optical amplifiers that replace the electrical regenerators that were previously used.

It is common knowledge that high capacity networks operate with signals in digital form. In this environment several optical systems have been proposed that can be distinguished both according the kind of modulation and detection. At the moment the optical systems that show the best properties in terms of simplicity, cost and robustness with respect to the degrading effects of a link, is the one based on an the Intensity Modulation with Direct Detection (IM-DD). For IM-DD systems the shape of the pulse is very important. Conventional IM-DD systems have been based on a rectangular shape, format known as Non Return to Zero (NRZ), but for high capacity transmissions it has been demonstrated that a Return to Zero (RZ) format can permit to increase the system capacity.

Optical fiber. The main limitations of the optical fibers are the chromatic dispersion or Group Velocity Dispersion (GVD) [1], the PMD [7], and the nonlinear effects [1].

The chromatic dispersion is a linear effect that induces a pulse broadening, generating intersymbol interference [1]. The PMD is an effect due to the fact that optical fibers for telecommunications have two polarization modes, even though they are called single-mode fibers. Such modes have two different group velocities that induce a pulse broadening depending on the input state of polarization of the signal [7].

The main nonlinear effects arising in single-mode fibers are Brillouin scattering, Raman scattering and Kerr effect [1].

Brillouin scattering is a backward scattering from acoustic waves, which can generate forward noise at the receiver. It is the lowest threshold fiber nonlinearity, but fortunately it can be easily suppressed in the most part of optical communication systems.

Raman scattering is a forward scattering from silica molecules. Raman response is characterized by a low gain and broad bandwidth (about 5THz) curve. This feature can severely limit the performance of WDM systems, by transferring the energy from one channel to the neighboring channels, if the power channel is too high.

Kerr effect is due to the dependence of the refractive index on the field intensity. It mainly manifests as a generation of new frequencies. When a single-channel signal is taken into consideration the Kerr effects induces a spectral broadening and the phase of the signal is modulated according to its power profile [1]. Kerr nonlinearity does not always show a negative impact, in fact in the regime of anomalous dispersion it induces a chirp effect that can compensate the degradation induced by the GVD. Such a compensation can be total if soliton signals are used [1][8]. When multichannel WDM signals are considered the Kerr effect can be more degrading since it induces both a nonlinear crosstalk among the channels that is called Cross Phase Modulation (XPM) and the generation of new frequencies, effect that it is called Four Wave Mixing (FWM).

Neglecting Brillouin and Raman effect, the signal propagation has been simulated by solving the nonlinear Schroendiger equation, that describes the signal propagation in optical fiber, by means of the split-step method.

Optical amplifiers. Two types of optical amplifiers have been considered in this work: EDFAs and SOAs. The behaviour of these amplifiers is quite different: gain saturation is a slow process in EDFAs while it is a fast process in SOAs. As a consequence a static model can be adopted in the first case while amplifier dynamics must be taken into account in the second. Optical amplifiers provide optical gain and add Amplified Spontaneous Emission (ASE) noise. In the case of EDFAs the optical gain can be assumed constant. On the contrary, in the case of semiconductor amplifiers, the instantaneous gain depends on the particular transmitted message and it is found by solving the rate equations [6].

Receiver and evaluation of system performance. In a common receiver the optical signal is transformed in an electrical current by a PIN photodiode, and electrically filtered. The signal timing is recovered by a baseband PLL and a threshold decision device decides the received bit [6]. The adoption of an optical EDFA preamplifiers brings the sensitivity of direct detection receivers quite close to the theoretical limit. After the optical preamplifier and before the photodiode, an optical Fabry-Perot filter is generically located to limit the presence of the ASE noise.

In the simulation code, the PIN photodiode and the electrical front end are simulated by a square law device and a Gaussian noise source that introduces the receiver noise. The electrical filter at the receiver is a second order Butterworth filter whose bandwidth is 0.8R.

The system performance are evaluated in terms of Q factor and time jitter [6][10] since some theoretical works [10], confirmed by experimental demonstration [4], have shown that an error probability lower than 10^{-9} can be achieved with a good approximation if the Q factor is higher than 6 and the standard deviation of the time jitter is lower than 6% of the bit time. This is the condition that we have chosen for the correct operation of an optical system.

2. METHODS TO OBTAIN THE HIGHEST CAPACITIES. The maximum bit-rate that can be achieved in an optical link for a single-channel system depends on several parameters and one of the most important is the GVD distribution along the propagation direction.

According to the GVD distribution we distinguish between regimes of propagation with high and low chromatic dispersion. We assume a regime with low chromatic dispersion when the only distortion induced by the GVD on the signal at the link output can be considered negligible and it is verified when the condition $R^2\overline{\beta_2}\,z \ll 1$ is satisfied, where R is the signal bit rate, z the link length, β_2 the GVD while the term $\overline{*}$ means the average along the link. It has to be pointed out that in such a propagation regime the system performance are not always good, since other effects can degrade the signal propagation as for example the nonlinear effects. Conversely the regime of high chromatic dispersion is when the signal propagation at the output link is deeply modified by the GVD and it manifests when $R^2\overline{\beta_2}\,z>1$. In the normal dispersion region this condition induces a large pulse broadening so preventing a correct system working. In the anomalous region the GVD induced chirp can be partially or even totally compensated by Kerr effect [1], allowing transmission with good performances.

When the local GVD is very low the Kerr effect can induce several signal degradations. Such a degradation can be very detrimental in the propagation of single-channel signals in the presence of ASE noise and in the case of WDM systems. In the first case the nonlinear interaction between signal and ASE induces a large spectral broadening when the link length is very long, conversely in the latter case also for short distance the FWM can induce a strong nonlinear crosstalk among the channels.

To limit the FWM it is preferable to operate with a local high GVD that is periodically compensated by devices having an opposite sign of the GVD. Such a devices can be simply optical fibers with opportune GVD and such a method is commonly called *dispersion management* [9]. In such a way the accumulated GVD can be very low and the FWM effect strongly limited. The dispersion management is the method that permits to achieve the highest capacity both using RZ and NRZ signal formats [5][9].

To obtain a satisfactory propagation of NRZ signals the overall link dispersion has to be kept very close to zero while, to efficiently propagate solitons, a small amount of anomalous chromatic dispersion is useful.

In the regime of high GVD the signal degradation can be limited by adopting several methods as either the dispersion management or the nonlinear compensation of the GVD by means of the Kerr effect [1]. The nonlinear compensation of the GVD is maximum when the pulse has the shape of an hyperbolic sechant and the peak power satisfies the following relationship:

$$P_k = \frac{\alpha\, L_A\, G\, |\beta_2|}{(G-1)\,\gamma\, \tau_s^2} \qquad (1)$$

where α is the fiber loss, G the amplifier gain and γ the nonlinear coefficient [1][8]; such a pulse is called *soliton*. Soliton signals seem the best pulse shapes to achieve the highest performance, however some effects can limit their propagation condition; such effects are the fiber losses, the nonlinear interactions and the Gordon-Haus effect [8]. The performance of soliton systems can be deeply improved by adopting in-line optical filters [5][8].

A comparison among the performances of different systems operating in DS fibers is presented in figure 1, where the maximum system capacity is reported versus the link length with a dispersion parameter equal to -1.28 ps^2/km and L_A=40 km. The maximum capacity has been obtained by optimizing the system parameters (mainly the input power and the time duration of the pulses) and increasing the bit rate until reaching the condition on the Q and on the time jitter are both satisfied.

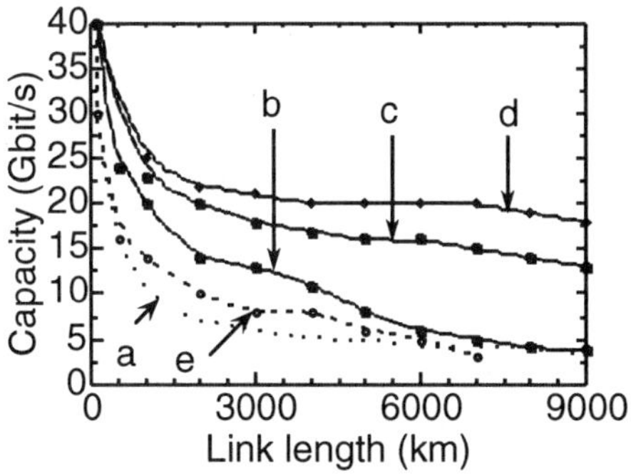

Figure 1: maximum capacity vs link length for links with DS fiber.

In figure 1 the different curves refer to the following systems

a) dispersion limited NRZ transmission,

b) soliton transmission in absence of in-line filters,

c) soliton transmission with fixed filters,

d) soliton transmission with sliding filters,

e) NRZ transmission with nonlinear compensation of the GVD without in-line filters.

All the curves show a decreasing of the capacity as the link length increases and soliton signals always have higher capacity with respect to NRZ signals. Some considerations on the curves (b) of fig. 1 deserve to be made. For very short link length the maximum capacity can be very high since all the degrading effects on the transmission are very weak. For z<3000 km, in the case of soliton signals, the main limiting effect is the soliton interaction; in fact the optimized time duration of the pulses remains almost constant passing from τ_s=5 ps for z= 40 km to 7 ps, while the time distance must grow as a function of the length and it induces a decreasing of the maximum capacity. For distances longer than 3000 km soliton interactions and Gordon-Haus effect become dominant.

In presence of in-line filters the system capacity deeply increases, especially in presence of sliding filters. The capacity of soliton systems is always higher with respect to the case of NRZ.

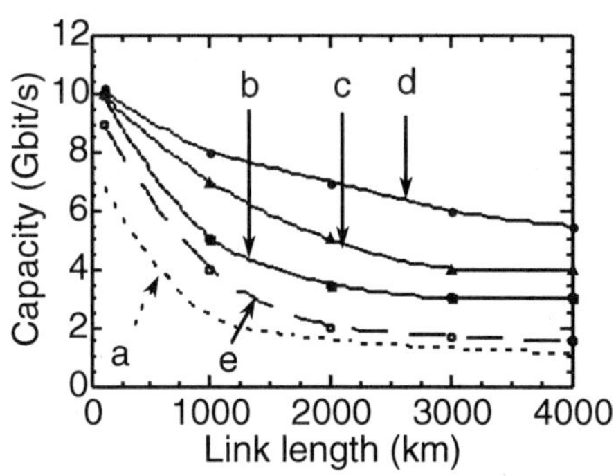

Figure 2: maximum capacity vs link length for links with step-index fiber.

In figure 2 the curves have been obtained for step-index fiber links with a dispersion parameter of -20 ps^2/km. In this case, due to high value of the GVD, the curves are reported up to a maximum length of 4000 km.

The different curves refer to the following systems

a) dispersion limited NRZ transmission,

b) soliton transmission in absence of in-line filters,

c) soliton transmission with fixed filters,

d) soliton transmission with sliding filters,

e) NRZ transmission with nonlinear compensation of the GVD.

The limitation due to the use of the step-index fibers is clearly shown by the comparison between figs.1 and 2.

A method to increase the system capacity is the use of the so called dispersion management. Generally, a good dispersion management is obtained when the product $\beta_2 L_L$ is high enough to reduce the FWM efficiency, but not so high to cause sensible interplay between Kerr effect and GVD.

In figure 3 we report the maximum capacity versus the link length for systems adopting dispersion management realized with only two different fibers.

The reported curves are evaluated in the following conditions

(a) refers to an NRZ system with a constant zero GVD;

(b) refers to an NRZ system using DS fibers and step-index fibers for dispersion compensation, in this case $\bar{\beta}$=0;

(c) refers to an NRZ system using step index fibers and DCFs, even in this case $\bar{\beta}$=0;

(d) refers to a soliton system using step index fibers and DCFs without in-line filtering, in this case $\bar{\beta}$=-1.28 ps^2/km;

(e) refers to a solitons system using step index fibers and DCFs with sliding filters, even in this case $\bar{\beta}$=-1.28 ps^2/km;

(f) refers to a soliton system using DS fibers and step-index fibers without in-line filtering, in this case $\bar{\beta}$=-0.2 ps^2/km;

The position of the compensating fibers was chosen according the results of ref. [9] that showed as in the case of NRZ signals it is better to locate fibers with normal dispersion at the output of the optical amplifiers to avoid modulation instability effects, conversely in the case of the solitons, at the output of the optical amplifiers it is better to locate fibers with anomalous dispersion. The period of the compensation is equal to 80 km in the case of links encompassing DS fibers and 40 for links encompassing step-index fibers.

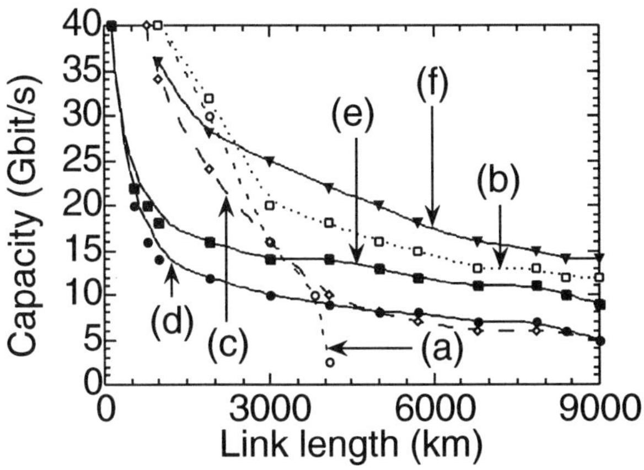

Figure 3: Maximum capacity versus link length for systems adopting dispersion management realized using only two different fibers. The details of the curves are reported in the text.

Curve (a) shows how in the case of the zero constant GVD the nonlinear interaction between signal and ASE is very strong and it deeply reduces the performance for distance longer than 3000 km. For longer distance the dispersion management is required as shown by the curve (b) and c. Curve (a) shows that when dispersion management on DS fibers is adopted, 10 Gbit/s transmission of NRZ signals up to 9000 km is allowed. By comparing the curves (c-e), obtained with dispersion management on step index fibers, for short distances NRZ systems outperform soliton ones while for long distances soliton systems show better performance, particularly when sliding filters are used. However, in cases (d) and (e), higher capacities can be achieved by increasing the average GVD (for an instance around -0.2 ps^2/km), and in particular for distances shorter than 3000 km capacities very similar to the case c can be obtained.

Soliton signals in links with dispersion management adopting DS fibers and compensation with step-index fibers (f) allows the system to achieve the highest capacity for distances longer than 3000 km, also in absence of sliding filters. This is due to the low average GVD that limits the nonlinear interaction, the soliton instability and the Gordon-Haus effect; furthermore the dispersion management permits to operate with higher input powers that permit to achieve higher signal-to-noise ratio and to reduce the time jitter with respect to the case with constant GVD. For distance shorter than 3000 km the capacity is a bit lower with respect to the NRZ signals since, also if low, the residual average GVD induces a stronger limitation with respect to the case with zero average GVD. It has to be pointed out, that for a GVD tending to zero, NRZ should be preferred since the nonlinear interaction between signal and ASE for RZ signal is worse and it is due to the higher input power required by the RZ signals with respect to the NRZ to obtain the same SNR.

Our results show that when dispersion management is used, single-channel soliton systems permit to achieve higher capacities with respect to NRZ on long distances and such capacities can be improved if in-line sliding filters are adopted.

Dispersion management is a fundamental method to increase the capacity also in WDM systems since it permits the reduction of the FWM effect and to increase the capacity of the channel bit rate.

Systems based on the Wavelength Division Multiplexing (WDM) permits to increase the capacity with respect to single-channel systems. Also for WDM systems the system performance and in particular the channeling characteristics depends on the GVD distribution. For an example in links encompassing step-index fibers it is preferable the use of a systems with many channels at low bit rate (622 Mbit/s or 2.5 Gbit/s), while in the case of DS fibers a lower number of channels but with higher bit rate (2.5, 5 and 10 Gbit/s) is preferable [5]. In both the cases the use of the dispersion management permits a large increasing of capacity especially if high channel bit rates are required.

Also the choice of the signal format depends on several links parameters. As shown by the simulations soliton WDM systems allows us to better exploit the fiber capacity, but also NRZ signals offer very high potentiality. The better solution seems the one of using different signal formats, both soliton and NRZ, within of the same optical bandwidth, depending on the value of the GVD in which the channel propagates [11].

4. GUIDELINES FOR IMPLEMENTATION OF HIGH CAPACITY SYSTEMS

From the results obtained by the simulations we can present some guidelines for the implementation of optical systems. We distinguish several classes depending on the link length: Short distances (z<500 km), medium distance (500 km<z<2000 km) and long distance (z>2000 km).

Short distance.

DS fibers. For distances up to 500 km single-channel capacities of 20 Gbit/s can be achieved both with NRZ and soliton signals. For soliton signal in-line filters are not required. By means of the dispersion management 40 Gbit/s can be reached both with solitons and with NRZ signals. By means of the WDM techniques, assuming particular channel bit-rates and frequency spacing, capacities up to 200 Gbit/s can be reached. Such WDM capacities can be increased by using compensation techniques to equalize the gain of the optical amplifiers.

Step-index fibers. Operating with erbium amplifiers at the wavelength of 1.5 μm, in the case of single-channel systems the capacity is lower of a factor equal to 4 with respect to the systems operating with DS fibers. The dispersion management is fundamental to transmit signal at 10 Gbit/s both by using solitons and NRZ signals. By means of the WDM technique the same capacities of the DS fibers can be reached, even though the channel bit rate of 10 Gbit/s can be used only in the presence of dispersion management.

Operating with SOA at 1.3 μm, for distance up to 500 km single-channel capacity around 15 Gbit/s can be achieved both with solitons and NRZ signals without dispersion management.

Medium distance

DS fibers. For a link up to 2000 km, single-channel system capacities of 20 Gbit/s can be obtained for NRZ signals by or operating with a GVD close to zero either by using dispersion management. For solitons such a performance can be obtained by using either in-line filters, or dispersion management, or with alternate polarizations [4] or using a constant GVD lower than 0.5 ps/nm/km. Conversely 10 Gbit/s can be obtained without any great complexities for soliton signals, while for NRZ signals the operation around the zero dispersion wavelength is required. By means of the WDM techniques a total capacity of 100 Gbit/s can be obtained, but limitations on the granularity are present and in particular only channel bit rate at 2.5 and 10 Gbit/s should be used. Using the same channel bit rate and assuming compensation techniques to equalize the gain of the optical amplifiers, total capacities of 200 Gbit/s can be reached.

Step-index fibers. Operating with erbium amplifiers at the wavelength of 1.5 μm, the dispersion management is fundamental to transmit signal at 10 Gbit/s both by using solitons and NRZ signals, conversely several controls on the average GVD and on the periodical compensation are required to have 20 Gbit/s. By means of the WDM technique the same capacities of the DS fibers can be reached, using channel bit rate down to 155 Mbit/s, while channel bit rate of 10 Gbit/s can be used only in the presence of dispersion management.

For these distance SOA links show more severe limitations with respect to links with erbium amplifiers, due to the strong saturation effects and to high power of the ASE noise. 10 NRZ signals can be transmitted up to 800 km, while soliton signals up to 2000 km if sliding filters are used.

Long distance

DS fibers. 10 Gbit/s NRZ signals can be transmitted only by means of dispersion managements, while for solitons no particular requirements is necessary but a right choice of the pulse duration and of the transmitted power. 20 Gbit/s can be reached by means of solitons signals in presence of dispersion management, or in-line filtering. The WDM techniques permit to reach capacities around 100 Gbit/s by using NRZ signals with unequally frequency spacing in the case of constant GVD and equally spacing with dispersion management. A bit higher capacity can be reached by means of solitons with in-line filters in links with constant GVD and without in-line filters in links with dispersion management. Higher capacities can be obtained by using the compensation techniques to equalize the gain of the optical amplifiers.

Step-index fibers. Operating with erbium amplifiers at the wavelength of 1.5 μm, in the case of single-channel systems 10 Gbit/s can be obtained with an accurate choice of the link parameters if dispersion management is used. Soliton signals with dispersion management permit a much better performance, especially if in-line filters are used or by adopting alternate polarizations. In the case of WDM systems the same capacities of DS fibers can be obtained if channel bit rate smaller of a factor equal to four are considered.

SOA links are not suggested for these distances.

Considerations on the capacities of transport networks. The numerical method used for the study of point-to-point links can be adopted also for the performance evaluation of signals propagating in path encompassing ADMs, OXCs and WCs. In many cases, for these devices, their transfer function and the noise contribution can be evaluated without too difficulty and as a consequence they can be introduced in our simulation code. Preliminary results have shown that 8 channels at 2.5 Gbit/s can be transmitted in networks with path 1000 km long having either OXCs or ADMs located every 100 km.

Acknowledgments. Work carried out in the framework of the ACTS ESTHER and UPGRADE projects. Fruitful discussion with the COST 239 participants are acknowledged.

REFERENCES

[1] G. P. Agrawal, 'Fiber-Optic Communication Systems' John Wiley & Son, INC, 1992.

[2] L. F. Mollenauer, P. V. Mamyshev, M. J. Neubelt, Optics Letters, **19**, 704, 1994.

[3] M. Nakazawa, et al. Electronics Letters, **31**, 565,1995.

[4] F. Favre, D. Le Guen, Electronic Letters, **31**, 991, 1995.

[5] F. Matera, M. Settembre, Journal of Lightwave Technology, **14**, 1 (1996).

[6] M. Settembre et al, J. of Lightwave Technology **15**, 962 (1997).

[7] F. Matera, C. Someda, "Anisotropic and Nonlinear Optical Waveguides" 1992 Elsevier Science Pubblishers

[8] A. Hasegawa, Y. Kodama, "Solitons in Optical Communications" *Clarendon press*, Oxoford 1995.

[9] F. M. Knox, W. Forysiak, N. Doran, J. of Lightwave Technology **13**, 1955 (1995).

[10] J. V.Wright, S. F. Carter, proc. of *Nonlinear Guide-Wave Phenomena*, Cambridge (UK), 1991, pp. 6-9.

[11] N. S. Bergano, et al. proc. of OFC'97, Dallas (Texas) 16-21 February 1997, PD21

PART SIX

Traffic modeling

Self-similarity in wide-area network traffic

Marco Isopi
Dipartimento di Matematica, Università di Roma, "la Sapienza"
P.le Aldo Moro 5 00161 Roma, Italy, isopi@mat.uniroma1.it

Abstract
Recent statistical work on traffic measurements are showing traditional models to be poor descriptors of what really goes on for WAN traffic. Here we discuss briefly the shortcoming of those models and illustrate some theoretical features of possibly more realistic models.

Keywords
Stochastic processes, network traffic, modelling

1 INTRODUCTION

Stochastic models of traffic in telecommunications systems are an important application of the modern theory of probability and stochastic processes. In turn problems coming from telecommunications have been stimulating the development of new mathemics at least since the work of Erlang. Queuing theory plays a central role, but approximation, limit theorems and most parts of the theory of stochastic processes are relevant in modelling and understanding traffic phenomema.

In the following we mention briefly some of the classical tools used in modelling and analysis and point at their shortcomings in describing what goes on in systems such as optic networks.

2 THE CLASSICAL TOOLS

The three basic ingredients of the conceptual framework are the Poisson process, Brownian Motion and the $M/M/1$ queue. Most of our intuition about stochatic behaviour is built by looking at this examples.

2.1 The Poisson Process

It is very convenient to think that some system is described by either a renewal process or a Markov process. Both the renewal and the Markov property tell you that the system will not be remembering for very long what its past states were; therefore, to describe its future, you don't need to look too much at its past history.

The Poisson Process is the only process that enjoys both the renewal and the Markov property. Here is a few features that make it extremely attactive as a paradigm in building models:

- The interarrival times T_i with exponential distribution

$$P(T_i > t) = e^{-\lambda t}$$

- Number of arrivals with Poisson distribution

$$P(N(t)) = e^{-\lambda t}\frac{(\lambda t)^k}{k!}$$

- The exponential distribution is memoriless
- For the Poisson Process you can compute almost everything explicitly
- Most traditional models for sources (e.g. renewal type models) share the same qualitative features.

2.2 The $M/M/1$ queue

The $M/M/1$ queue is built with a couple of Poisson processes. One describes customers arriving at a shop and the other the owner of the shop taking care of the customers with a 'first come, first served' policy. Almost anything we want to know about the $M/M/1$ queue can be computed explicitly. (see e.g. (Cooper 1981))

2.3 Brownian Motion

The ubiquitous nature of Brownian Motion is due to the Central Limit Theorem. In most applications it is fair to say that, when suitably rescaled, *every process tends to Brownian Motion.* You only need some reasonable decay of correlations and marginal distributions satisfing very mild conditions. In particular almost everything we can build by toying with Poisson processes will be well approximated by Brownian Motion. A Brownian Motion X_t is characterized by the following properties:

- X_t has stationary indipendent increments
- X_t has gaussian distribution with mean 0 and variance $\sigma^2 t$
- X_t has continuos trajectories

2.4 Classical models

For several years we have been using a wide variety of stocastic models to describe traffic. All of this models share the basic qualitative features of at least one of the three basic examples mentioned above.

We may not be able to compute explicitly as many relevant quantities as we can get in our three basic examples, but we know that the qualitative behaviour is not radically different and our intuitions works well as a guide for getting quantitative estimates of what goes on.

We will not even try to list these models here, but refer the reader to the recent excellent review (Jagerman *et al.* 1997).

3 NOVELTIES

Recent traffic measurement seem to fit poorly with the clasical tools we described so far. They display unfamiliar statistical features. Here are the most apparent ones:

- Heavy tailed distributions (with infinite variance)
- Long Range Dependence
- Self-similarity

3.1 Heavy-tailed distributions

The random variable X is said to have a heavy-tailed distribution F if

$$1 - F(x) = P(X > x) \sim x^{-\alpha} L(x)$$

where $L(x)$ is slowly varying at infinity. (see (Samorodnitsky *et al.* 1994))

The expectation of a heavy-tailed random variable need not be finite: $E[X^\beta] < \infty$ if $\beta \leq \alpha$ and $E[X^\beta] = \infty$ if $\beta > \alpha$.

A case of special interest is $1 \leq \alpha < 2$, the so called Noah effect (finite mean, infinite variance, see (Mandelbrot 1982)). Teletraffic data presents statistical evidence of heavy tailed distributions (see (Resnick 1997)).

3.2 Long-range dependence

We say a process $X = (X_i \ , \ i = 1, 2, \ldots)$ is covariance stationary if $\rho(i,j) = \mathrm{Cov}(X_i, X_j) = \rho(|i-j|)$.

A stationary process is said to exhibit long-range dependence (long memory, *Joseph effect*, persistence) if its covariance decays slowly

$$\rho(k) \sim k^{2H-2} \quad k \to \infty \quad \frac{1}{2} < H < 1$$

As a measure of long-range dependence we used the Hurst parameter H, with $\frac{1}{2} \leq H < 1$. Processes with long range dependance are notoriously hard to analyze, but there are effective techniques available ((Beran 1994), (Cox 1984)).

3.3 Self-similarity

We say a stationary process X is **exactly self-similar** (with self-similarity parameter H) if for all $m = 1, 2, \ldots$

$$X \approx m^{1-H} X(m) \quad \text{(in distribution)}$$

where for $k > 0$,

$$X^{(m)}(k) = m^{-1}(X_{km-m+1} + \ldots + X_{km})$$

X is **asymptotically self-similar** (with self-similarity parameter H) if

$$X \approx m^{1-H} X(m) \quad \text{as } m \to \infty$$

If we are restricting ourselves to a second order description of a process (which as you see may well not be sufficient!), we can similarly define exact/asymptotic second-order self-similarity:

$$\mathrm{Var}(X^{(m)}) = \sigma^2 m^{2H-2} \quad \text{for all } m \geq 1$$
$$\mathrm{Var}(X^{(m)}) \sim \sigma^2 m^{2H-2} \quad \text{as } m \to \infty$$

Self-similar processes exhibit fractal-like behaviour; their graph tends to look the same when you look at it from different distances. Therefore they capture well the idea that all time scales are relevant in the description of the process. The three figures in this tutorial are taken from ((Pagano 1998)) and

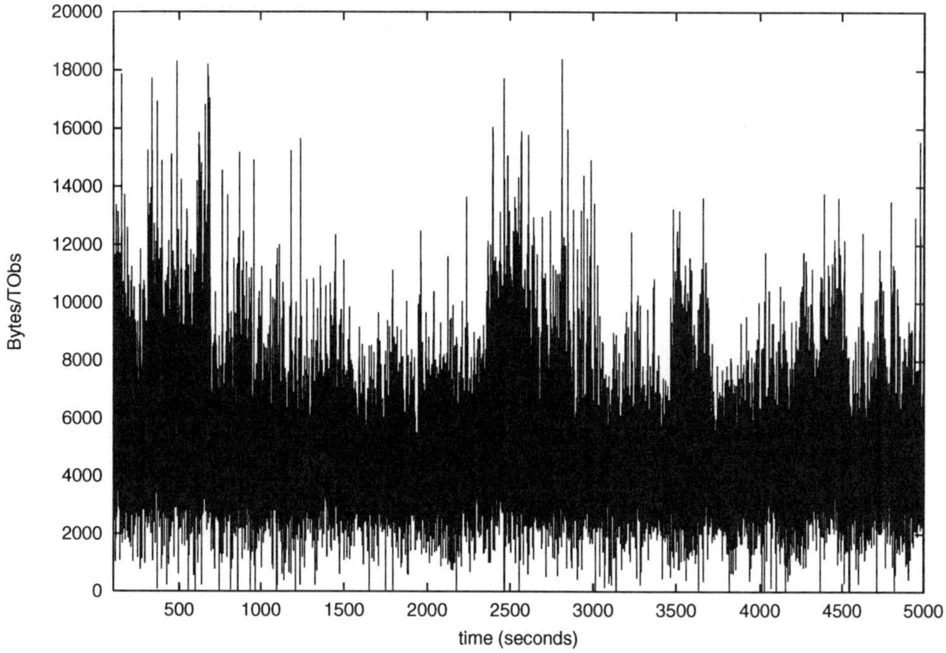

Figure 1 LAN traffic observed every 0.1 s. T_{obs}=1 s

show actual measures of LAN traffic on three different time scales. Anything which is well described by one of the traditional models should become flatter and flatter as you look at it on larger time scales (this is just a consequence of the law of large numbers).

While you can reproduce this data's qualitative and quantitative features with a parsimonious self-similar model, traditional models perform badly. You need a large number of parameters to fit the data and, as traces get longer, you find you need quickly more and more.

If we *assume* the process to be self-similar, then it is easy to extrapolate (a longer session is *not* simulated by glueing together shorter sessions) *and* interpolate.

3.4 Self-similarity in LAN and MAN/WAN traffic

There is increasing experimental evidence for self-similarity in LAN traffic (see figures 1, 2 and 3 , (Leland *et al.* 1994))

We can identify at least two causes for this. A simple fact is that single sources exhibit strong LRD. Less immediate, but of great importance is the fact that the Noah effect leads to the Joseph effect: the superposition of a

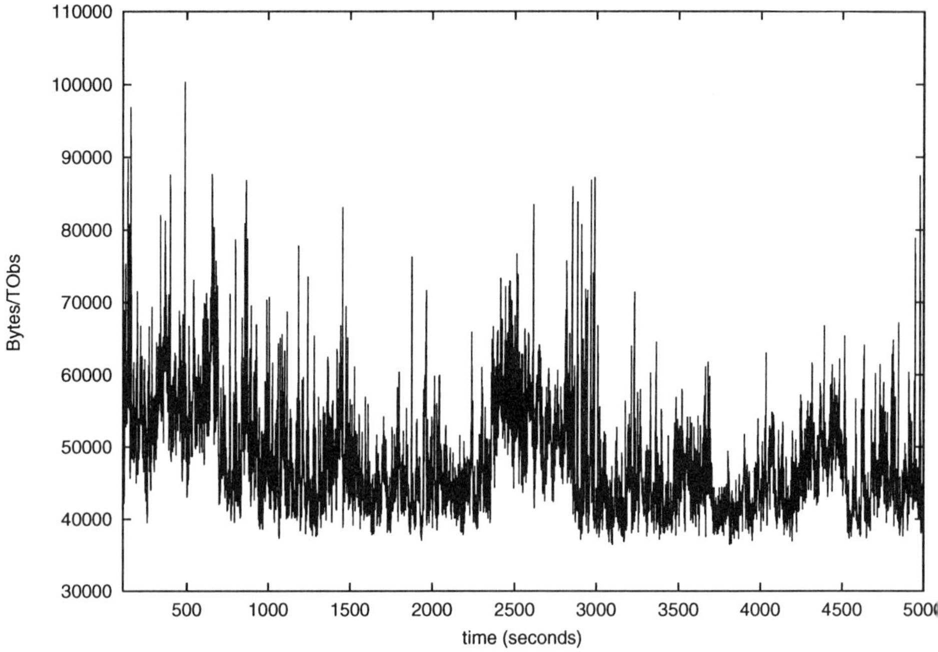

Figure 2 LAN traffic observed every 1 s. T_{obs}=1 s

large number of ON-OFF sources with heavy-tailed ON or OFF periods gives rise to Fractional Brownian Motion. (see (Willinger 1995))

More appropriate for WAN traffic modelling is Kurtz's theorem: consider a Poisson number of source activations with the holding time of each source heavy-tailed with infinite variance. Then, the normalized workload converges to FBM. As noted by Willinger and other authors one finds statistical evidence of heavy tails and self-similarity in single components of the Internet and other WANs.

- File system events are self-similar
- CPU time of a typical UNIX process is heavy-tailed with infinite variance
- File sizes on file servers and document sizes on web servers are heavy-tailed with infinite variance
- TCP and HTTP connections sizes/durations are heavy-tailed with infinite variance

It is not still clear what are all the consequences of self-similarity, but we can list a few here ((Erramilli *et al.* 1997)):

- Queue lenght distribution no longer exponential

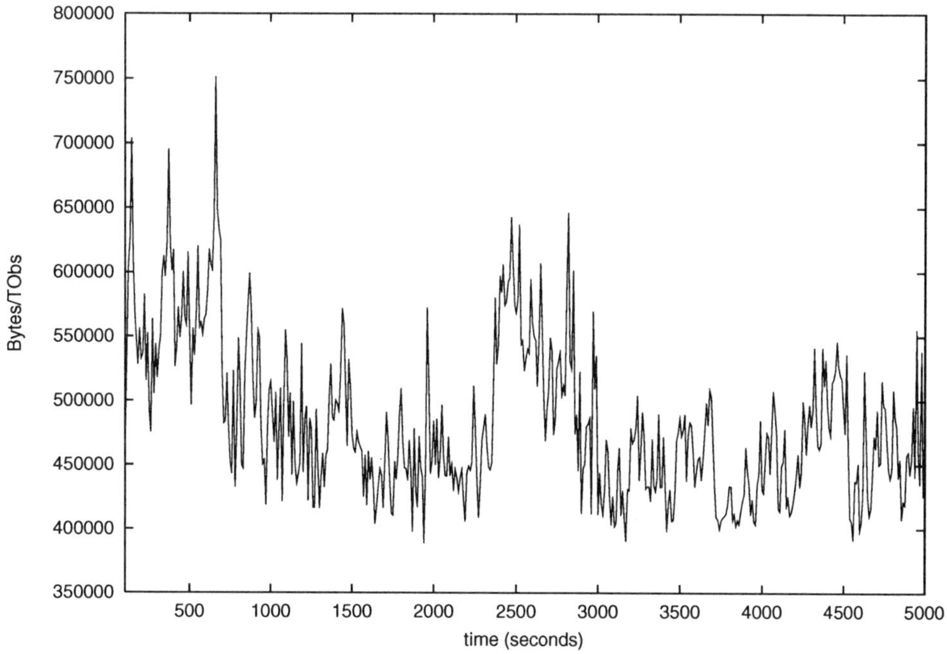

Figure 3 LAN traffic observed every 10 s. T_{obs}=10 s

- Very different (worst) heavy traffic behaviour
- No significant expected gain from within source multiplexing
- Potential for multiplexing gains across sources

4 CONLUSION AND SOME OPEN ISSUES

These new features cannot be ignored, but they do not mean necessarily bad news. Certainly they leave a number of open issues. The main one is of course how to exploit self-similarity. On a more particular level there is still a lot of work to do on fast generation of fractal traffic for simulation and the analysis of networks of queues. Lastly we notice that LRD and the fact that one expects no gains from within source multiplexing, mean that we need to rethink the structure of tariffs and devise a rational pricing scheme for self-similar or bursty traffic (see (Kelly 1997)).

In conclusion I can say that from the point of view of an applied mathematician these new finds in traffic data open are opening the way for a lot of fruitful work in the years to come.

REFERENCES

Beran, J. (1994) *Statistics for Long-Memory Processes.* Chapman & Hall, New York.

Çinlar, E. (1975) *Introduction to Stochastic Processes.* Prentice–Hall, Englewood Cliffs.

Cooper, R.B. (1981) *Introduction to Queuing Theories.* North Holland, New York.

Cox, D.R. (1984) Long–range dependence: A review, in *Statistics: An Appraisal* (ed. H.A. David and H.T. David). The Iowa State University Press, Ames.

Erramilli, A. , Narayan, O. and Willinger, W. (1997) Fractal queuing models, in *Frontiers in Queuing* (ed. J.H. Dshalalow). CRC Press, Boca Raton.

Jagerman, D.L. , Melamed, B. and Willinger, W. (1997) Stochastic modelling of traffic processes, in *Frontiers in Queuing* (ed. J.H. Dshalalow). CRC Press, Boca Raton.

Kelly, F.P. , (1997) Charging and accounting for bursty connections, in *Internet Economics* (ed. L.W. McKnight and J.P. Bailey,). MIT Press, Cambridge, MA.

Kurtz, T.G. (1996) Limit theorems for workload input models, in *Stochastic Networks: Theory and Applications* (ed. F.P. Kelly, S. Zachary and I. Ziedins). Claredon Pres, Oxford.

Leland, W.E. , Taqqu, M.S. , Willinger, W. and Wilson, D.V. , (1994) On the self-similar nature of network traffic (extended version). *IEEE/ACM Trans. on Networking,* **2**:1, 1–15.

Mandelbrot, B.B. (1982) *The Fractal Geometry of Nature.* W.H. Freeman and Co. , San Francisco.

Pagano, M. (1998) *Analisi Multiscala e Valutazione delle Prestazioni in Reti a Larga Banda.* Tesi di Dottorato, Università di Pisa.

Resnick, S.I. , (1997) Heavy tail modeling and teletraffic. *The Annals of Statistics,* **25**, 1805–1869.

Samorodnitsky, G. and Taqqu, M.S. (1994) *Stable non–Gaussian Random Processes: Stochastic Models with Infinite Variance.* Chapman & Hall, New York.

Willinger, W. , Taqqu, M.S. , Sherman, R. and Wilson, D.V. , (1995) Self-similarity through high-variability: statistical analysis of Ethernet LAN traffic at the source level (extended version). Preprint.

5 BIOGRAPHY

Marco Isopi received his PhD in Mathematics from the Courant Institute. He teaches Probability at the University of Rome "la Sapienza".

The impact of the Hurst parameter on queueing performances using chaotic sequences

Authors: E. Costamagna, L. Favalli, P. Gamba, G. Iacovoni [(*)]

Dipartimento di Elettronica, Università di Pavia, Italy

(*) Also with CORITEL, *Rome, Italy*

e-mail: giovanni@comel1.unipv.it

tel: 06-204110032- fax: 06-20410037

Abstract

The need to understand the nature of packet traffic is growing more and more as its volume is increasing, due to the development and the deployment of optical network technologies. The Hurst parameter is the key figure which captures the degree of Long Range Dependence in the new traffic paradigms. Its impact on network engineering is still unclear and object of a strong debate, especially after the discovery of the so-called "cross-over effect".

In order to give a contribution to this discussion, in this paper we compare the queueing simulation results when an infinite buffer is loaded with different sequences, each sequence having a different Hurst parameter value, but the same marginal distributions. Simulations have been carried out using different phase-shifts and load conditions.

Keywords

ATM, Hurst parameter, Long range dependence, Self-similarity

1 INTRODUCTION

The volume of traffic carried over telecommunication networks is increasing, following the development and the deployment of optical technologies. A significant part of this traffic is transported and switched using a packet-like structure. The need to understand the nature of *packet traffic* is growing more and more as it gives birth to difficult engineering problems such as bandwidth allocation and buffer sizing.

Through extensive measurements on working packet networks, the fractal-like characteristics (or self-similar properties) of aggregated packet streams, whatever the type of traffic (Ethernet's LANs, VBR video, ISDN control channel, etc.), has recently been shown. A key feature of self-similar processes is the "Hurst effect" or "persistence", i.e. the correlation between records observed over different time scales; this effect can be captured by the so called Hurst-parameter H.

More precisely, let $\{X_i\}$ be a covariance stationary process and let $\{X(m)_i\}$ be the aggregated process obtained by averaging the original series over non-overlapping intervals of size *m;* then there exists H (the Hurst parameter) such that:

$$\text{var}\,\{X(m)\} \sim m^{-2(1-H)} \qquad \text{as } m \rightarrow \infty \qquad (1/2 \leq H < 1) \qquad (1)$$

If H = 0.5 the process has a short-range dependence (SRD), whereas H > 0.5 characterizes a process with long range dependence (LRD), also called *asymptotically self-similar.* If $\text{var}\{X(m)\} = m^{-2(1-H)}$ is true for all $m \geq 1$, then the process is *exactly self-similar*, as for instance the Fractional Brownian Motion (FBM).

The Hurst parameter can therefore be put in relation with the behaviour of the process when it is considered its low frequencies spectra.

After the above mentioned discovery of fractal-like characteristics (which implicate LRD) in teletraffic, there has been a strong debate whether or not engineering formulae can still be based on traditional markovian models, known to be SRD. Someone claims [1] that the latters lead to underestimation of the bandwidth requirements, while other researches [2] show a succesful use of markovian models.

The discrepancy between the two above sets of seemingly contradictory results has been explained through the help of the "cross-over" effect: when a large enough number of i.i.d. sources are multiplexed, larger multiplexing gains are obtained with high-H sources than with low-H sources. More precisely, for FBM traffic, in[3] it has been shown that, given two FBM streams with the same mean μ and the same variance σ but with different Hurst parameters H_1 and H_2 then for a finite level B in an infinite buffer, there exists a value μ^* such that for $\mu \geq \mu^*$ the stream with the higher H requires smaller bandwidth to meet a specified overflow probability requirement.

Through simulations, the same cross-over effect was observed in [4] where the performances of VBR video traces (which are only *asymptotically self-similar*) were compared with those of a finite-state markovian model: a large number of multiplexed i.i.d. sources requires less bandwidth when each source has H = 0.7 than when it has H = 0.5.

In order to better understand the engineering significance of the H parameter, in section II of this paper we present the queueing simulation results for an *infinite* buffer loaded each time with a different sequence. These sequences are derived from a set of three traces; **each trace**

has a different H, but all of them exhibit the same marginal distribution, i.e. the number of packets in a GOP period (one GOP = 12 video frames, each one lasting 1/25 sec). One trace is a real VBR video, while the other two have been built using chaotic maps; in a previous work [5] we validated the modelling capability of the two chaotic traces by comparing the QoS parameters of an ATM multiplexer of *finite* size fed by the modelled traces with those resulting from the actual VBR video load.

In section III we briefly discuss the results of these simulations, emphasising the role played by the H parameter.

2 SIMULATION RESULTS

The VBR video trace is the cartoon ASTERIX, MPEG coded, with high activity [for a discussion of activity see 6], since H=0.93. The two chaotic traces (CAOS_1 and CAOS_2, for detail refer to [5]), built by sampling the trajectory of the Lorenz attractor, simulate ASTERIX at the Group of Picture Level (GOP). This modelling approach is similar to that described in [7], where chaotic mechanisms are used to simulate the physical process of channel error generation.

Some relevant statistical properties are summarized in the following table:

Table 1 Statistical properties of viedo trace

	Mean (kbit/sec)	σ (Kbit/sec)	H
MPEG	558	259	0.81
CAOS_1	558	259	0.68
CAOS_2	558	259	0.78

H values have been calculated using time-variance plots ([6]); this is just a statistical method which tries to estimate an asymptotic behavior from a finite data set. Therefore these values must be considered with care.

Each trace is packetized in ATM cells introducing AAL5 overhead (one AAL5 PDU made of 8 ATM cells). The cells are then evenly spaced over a GOP period; this, however, is not critical since it has been noted that only low frequencies in the power spectra dominate queueing behaviour [8].

In [5] we showed that when feeding *finite* buffers with deterministic service time, FIFO mode, with CAOS_2, the obtained QoS parameters were highly consistent with those obtained with the MPEG trace for buffer size up to 100 cells; we related this result to the good matching on the time variance plot of the two sequences for low *lags* (fig. 1). In the same way, CAOS_1, which matches the MPEG values for larger values of *lags* (fig. 2) gave good results for 10,000 cells buffer.

Now we extend the simulation of queueing performances to the three sequences feeding an *infinite* buffer. First, for the sequence $\{X_i\}$ corresponding to the MPEG trace we produce the new sequence:

$$Y_i = \sum_{j=1}^{j=6} X(i + nj)$$

For n = 10 this leads to the sequence MPEG_SHIFT10, while n = 100 leads to the sequence MPEG_SHIFT100; in other words we obtain two phase-shifted versions of the same MPEG stream. We give the following two figures in order to have an approximate idea about the degree of MPEG autocorrelation corresponding to these two phase-shifts: at 10 *lags* it is still remarkable (0.36), while at 100 *lags* it has reached quite a small value (0.1).
Similarly we produced on one hand CAOS_1_SHIFT10 and CAOS_1_SHIFT100, and on the other hand CAOS_2_SHIFT10 and CAOS_2_SHIFT100.

Each of the 9 sequences is sent to the infinite ATM buffer; the outgoing data frequency is varied so as to consider a load ranging from 10% to 100% (with 10% step size): this should correspond to varying the number of multiplexed streams when the service time is kept constant.

For phase-shift = 0, fig. 3a depicts the obtained mean queue length (mql) versus load for the three sequences; similarly fig. 3b and 3c show the mql for SHIFT10 and SHIFT100 respectively. Besides, for three values of load (50%, 70%, 100%), fig 4 and 5 show the complementary queue length distributions for the chaotic sequences compared to those of the actual MPEG traces, when no shift is applied. Similar results (not shown here for brevity) are obtained when considering the delay distributions.

3 DISCUSSION

Here are some observations derived from the results:

1) For loads up to 40% and greater than 80% **the chaotic sequences** exhibit a behaviour similar to the original traces. Besides, the chaotic sequence CAOS_1_NO_SHIFT mimic quite well the MPEG_NO_SHIFT sequence in fig. 4; similarly, the mean queue length of CAOS_1 and that of the MPEG sequences are almost always the same (fig. 3). This supports the feeling that the behavior of the chaotic sequences is well matched to the simulated traffic for a proper choice of the attractor parameters, load and phase-shift

2) **Cross-over effects** can actually be observed (see for instance fig. 4). However more crossing points between two curves (see for instance fig. 4a) can be noticed and, strikingly, cross-overs can take place several times as the load is increased. A better definition of cross-over effect is therefore needed

3) Apparently the **phase-shift** results in separating the chaotic curves from the original one, but only for average loads; the larger the phase-shift (and therefore the less the degree of correlation), the larger the distance. Anyway, the phase-shift has a considerable impact on cross-over effects, and it deserves further studies

4) **In general**, the chaotic sequences CAOS_2, differs with respect to queueing performance from the MPEG sequences, even if the H parameters are very similar (0.81 and 0.78): see fig. 3 and 5. On the other hand the CAOS_1 (H=0.68) sequence exhibits a closer behavior to that of the MPEG one (see for instance the result for the mean queue length, fig. 3). A reason for this could be found in the similarity of the *actual* values in time-variance plot for large lags, and not in the similarity of the H parameters.

Obviously, all these questions and these preliminary results need further investigation; therefore we intend to perform new simulations, with aggregated traffic and other scenarios (type of traces, other H values, etc.).

Thanks to Marco Isopi (University of Rome La Sapienza) for useful discussions.

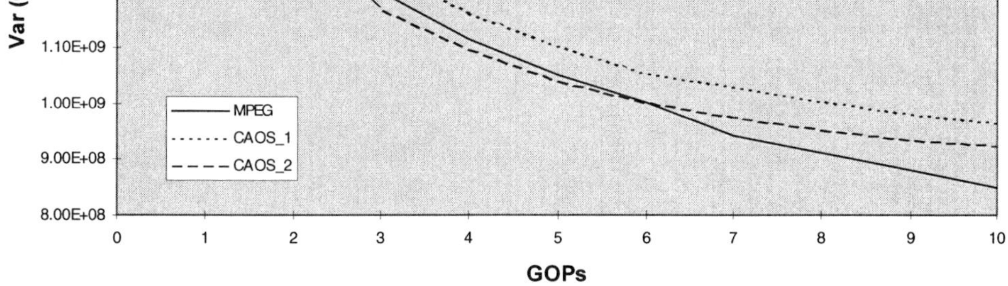

FIG.1. Variance-time plot (linear scale) in the range [0:10]

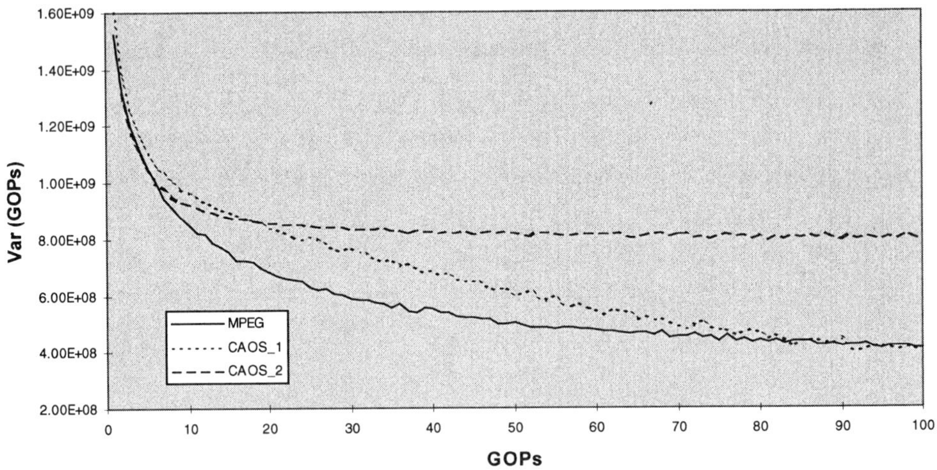

FIG.2. Variance-time plot (linear scale) in the range [0:100]

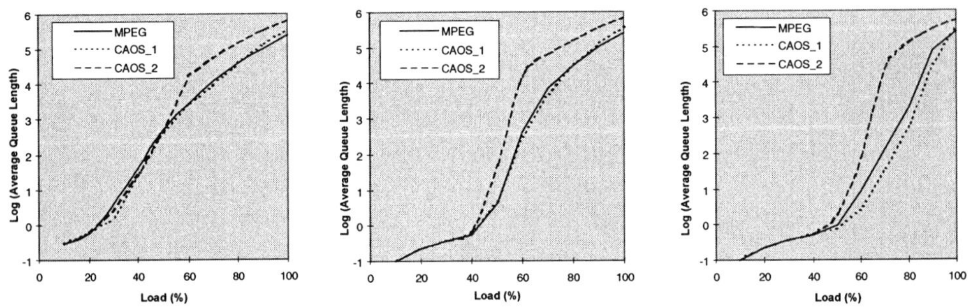

a) phase shift = 0 b) phase shift = 10 c) phase shift = 100

FIG.3. Average queue length

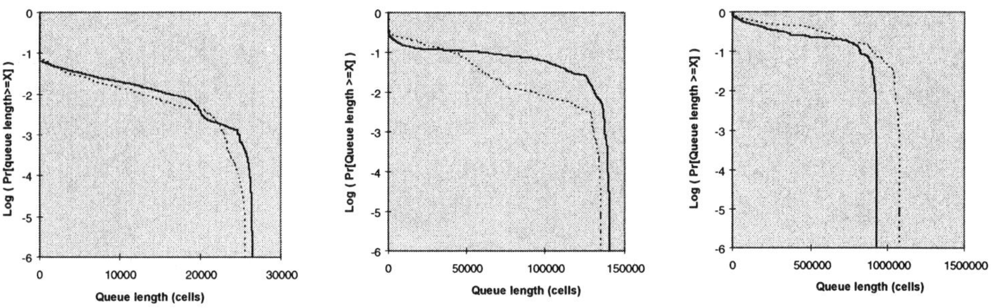

a) load = 50% b) load = 70% c) load = 100%

FIG.4. Complementary distribution function (MPEG: continuos line; CAOS_1: dotted line)

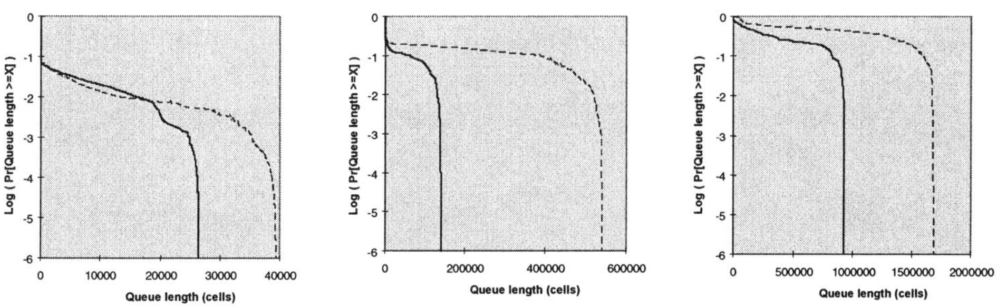

a) load = 50% b) load = 70% c) load = 100%

FIG.5. Complementary distribution function (MPEG: continuos line; CAOS_2: dashed line)

References

[1] A. Erramilli, O. Narayan, W. Willinger: "Experimental queueing analysis with long-range dependent packet traffic", *IEEE Tr. On Networking*, 4, 2, 1996

[2] D.P. Heyman, T.V. Lakshman, "What are the implications of long-range dependence for VBR-video traffic engineering?", *IEEE Tr. On Networking*, 4, 3, 1996

[3] K.R. Krishnan, "A new class of performance result for fractional brownian traffic model", *Queueing System* 22, 1996

[4] K.R. Krishnan, G. Meempat, "Long-range dependence in VBR video streams and ATM traffic

engenering", *Performance Evaluation*, 30, 1997

[5] E. Costamagna, L. Favalli, P. Gamba, G. Iacovoni: "A simple model for VBR traffic based on chaotic maps: validation through evaluation of ATM multiplexers QoS parameters", *IEEE-ICC '98*, accepted for publication

[6] J. Beran, R. Sherman, M.S. Taqqu, W. Willinger: "Long-range dependence in variable bit rate video traffic", *IEEE Tr. On Communications*, 43, 2/3/4, 1995

[7] G. Benelli, E. Costamagna, L. Favalli, P. Gamba: " Satellite channels modelled by chaotic bit error generators", *EMPS '96* ,Rome, Italy, Oct. 1996

[8] S. Li, J. D. Pruneski, "The linearity of low frequency traffic flow: an intrinsic property in queueing systems", *IEEE Tr. On Networking,* 5, 3, 1997

Performance Analysis of a Single Server Queue Loaded by Long Range Dependent Input Traffic

R.G. Garroppo, S. Giordano, M. Pagano, G. Procissi, F. Russo
Department of Information Engineering - University of Pisa
Via Diotisalvi 2 I-56126 Pisa - Italy
tel.+39 50 568539 fax +39 50 568522
e-mail: {garroppo@, giordano@, pagano@,
procissi@edison., russo@}iet.unipi.it

Abstract

Recent studies on broadband networks show that real traffics exhibit Long Range Dependence (LRD). Because of its dramatic consequences on queueing performances this property has to be taken into account in network design and management. To these aims we analysed the behaviour of a single server queueing system loaded by an LRD arrival process modelled by fractional Gaussian noise. Our goal was to obtain an analytical estimation of the complementary probability, that represents an upper bound for Cell Loss Ratio. The approach is based on the Large Deviations Theory and we successfully validated the results by means of simulation studies.

Keywords

Long Range Dependence, Large Deviations Theory, Queueing Performances

1 INTRODUCTION

The LRD feature represents a relevant statistical memory property of traffic flows in LANs, MANs, WANs, D-channel signaling in ISDN networks and VBR coded video sequences (Willinger, 1996). This characterising behaviour implies an hyperbolic decay of the traffic data autocorrelation that cannot be parsimoniously captured using traditional Markov processes.

The LRD correlation structure is related to the high burstiness presented by multimedia traffic. When investigating the temporal behaviour of traffic patterns, the eyeballing evidence of LRD is the presence at each time scale of clusters of arrivals, called *bursts*. Hence indexes of burstiness based on the *maximum burst length* concept have to be replaced by new metrics able to characterise the traffic behaviour at the different time scales. In this framework a promising approach is the introduction of the Hurst parameter H, which expresses the degree of self-similarity and represents a measure of LRD and therefore of the related burstiness: the greater is the burstiness of the traffic, the higher is its Hurst parameter.

In the future telecommunications scenario a major role will be played by broadband networks, based on ATM architecture. A key characteristic of these networks is the provision of Quality of Service (QoS) requirements. As a consequence it is particularly important to obtain some quantitative results on network performances in terms of cell loss ratio, maximum cell delay, cell delay variation, delay jitter etc.. The estimation of these parameters involves the study of elementary queueing systems, which represent simplified models of single network elements. The mentioned long memory property has a deep impact on queueing performances because of the bursty nature of the arrival process. Moreover, traffic models capturing LRD features makes unfeasible the analytical study of even very simple queueing systems.

In this paper we analysed an infinite buffer single server queue to estimate the complementary probability of buffer occupancy, that gives many useful insights on QoS parameters. In fact it is widely used as an upper bound for the estimation of the overflow probability for finite buffer systems. Typical loss probability requirements for broadband networks are around 10^{-6}-10^{-8}: to estimate these target values a simulation based approach becomes impassable, since it requires unacceptable long running times. To overcome these difficulties, two different techniques have been considered: rare event simulation procedures (Giordano, 1998) and mathematical approaches based on Large Deviations Theory (LDT).
In this work we have considered the latter philosophy, which leads to analytical results very useful for the design of connection admission control schemes. The paper is organised as follows: in section 2 we present an overview on LDT introducing basic concepts such as the rate function and the Large Deviations Principle. In section 3 we show some theoretical results achieved by the application of LDT to fractal queueing networks, while in section 4 we validate the analytical results concerning the overflow probability by means of discrete event simulations. In the Conclusions Section we highlight the efficiency of our approach and further research directions.

2 NOTES ON LARGE DEVIATIONS THEORY

In this section we introduce some concepts of LDT which are necessary in the paper (for a more complete analysis see (Dembo, 1993), (Shwartz, 1995) and (Bucklew, 1990)). The function $I: R \to (-\infty,+\infty]$ is assumed to be a *rate function* if it satisfies the following conditions:

- I is non-negative;
- I is lower semi-continuos;
- its *level sets* are compact (*good rate function).*

Moreover, the pair $\left\{\frac{W_t}{a_t}, v_t\right\}$ satisfies a *Large Deviations Principle (LDP)* with rate function I if the following conditions hold:

- For every closed subset $F \subset R$:

$$\limsup_{t\to\infty} \frac{1}{v_t} \log P\left(\frac{W_t}{a_t} \in F\right) \le -\inf_{x\in F} I(x) \tag{1}$$

- For every open subset $G \subset R$:

$$\liminf_{t\to\infty} \frac{1}{v_t} \log P\left(\frac{W_t}{a_t} \in G\right) \ge -\inf_{x\in G} I(x)\,. \tag{2}$$

Roughly speaking, if the pair $\left\{\frac{W_t}{a_t}, v_t\right\}$ satisfies an LDP then

$$P\left(\frac{W_t}{a_t} > x\right) \approx e^{-v_t I(x)}.$$

The previous expression justifies the name *rate function* given to I: in fact it establishes the decay rate for the tail of the mentioned random variable distribution.

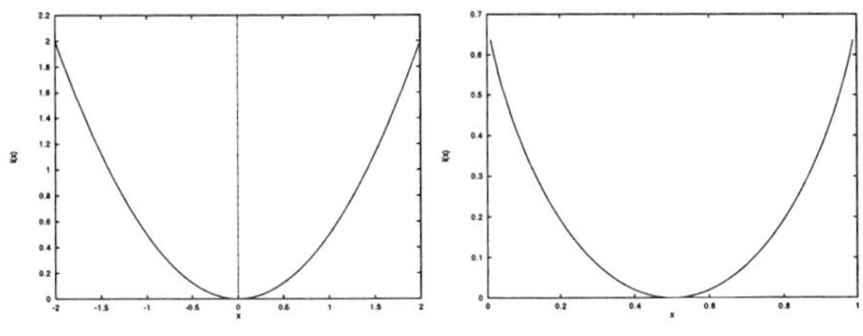

Figure 2. Rate functions for Gaussian standard (left) and Bernoulli (right) random variables.

For example, the sequence of empirical means $\{M_n\}$

$$M_n = \frac{X_1 + X_2 + \cdots + X_n}{n}$$

of i.i.d. random variables X_n satisfies a LDP (see *Cramér Theorem* in reference (Dembo, 1993)) with good rate function depending on the X_n distribution. In figures 1 e 2 we show the plots of I-functions for Gaussian, Bernoulli, Poisson and Exponential random variables; as can be seen I is usually convex.

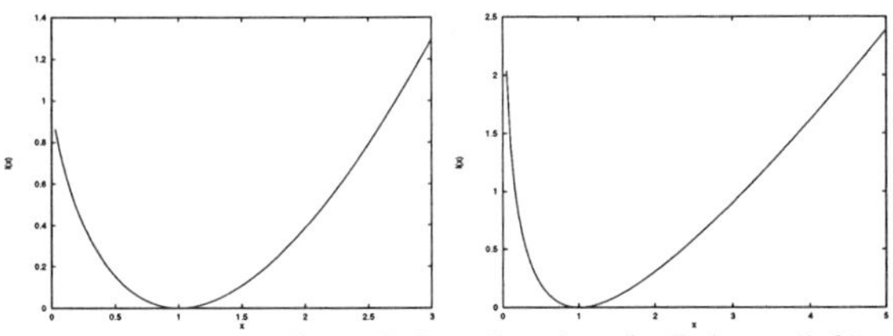

Figure 3. Rate functions for Poisson (left) and Exponential (right) random variables.

3 QUEUEING PERFORMANCES WITH LDT

A widely used queue model in ATM environment is represented by a single server queue with deterministic service rate C (because of the fixed-length of information units, i.e. cells).

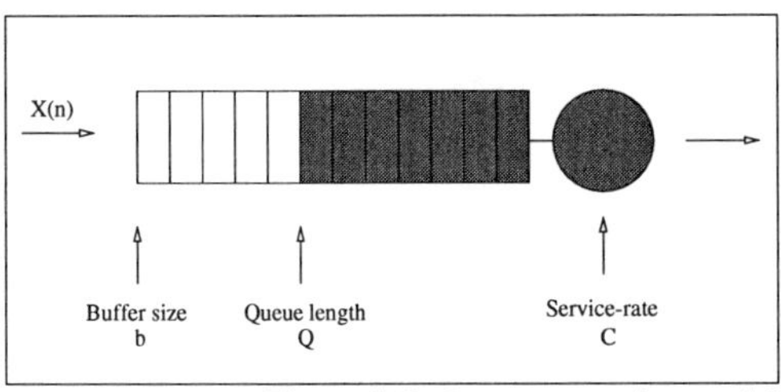

Figure 4. Reference queueing model.

In figure 3, we show the relevant parameters of the actual queueing model. In order to obtain an estimation of an upper bound for the overflow probability when the queue is fed by LRD traffic processes, we applied LDT to a simplified version of that model, characterised by an infinite buffer.
In the analysis we considered a slotted time *fractal queueing system*, whose input process X_t represents the number of arrivals in the time interval $((t-1)T_u, tT_u]$, where T_u is the time unit (assumed equal to one) and $t \in Z$.

The arrival process is modelled as follows:

$$X_t = m + G_t$$

where G_t is the so-called *fractional Gaussian Noise (fGn)* and m represents the mean arrival rate. The *fGn* is a discrete-time, stationary, Gaussian, exactly second-order self-similar (and so LRD) process with zero mean, Hurst parameter H and autocorrelation function:

$$r(k) = \frac{1}{2}\left(|k-1|^{2H} - 2|k|^{2H} + |k+1|^{2H}\right).$$

The associated workload process, representing the difference between the total amount of work offered to the queue in the interval *(0, t]* and the maximum number of served arrivals in the same time period, is given by:

$$W_t = B_t - (C-m)\cdot t$$

where B_t is the sample of a *fractional Brownian motion* process at time tT_w, which is related to G_t by relation $G_t = B_t - B_{t-1}$.

Denoting with Q the asymptotic queue length and considering the assumed weakly stationarity of arrival and service processes, from the Lindley equation (Duffield, 1994) it follows:

$$Q = \sup_t W_t$$

Substituting a_t and v_t with t and t^2/σ_t^2 (where σ_t^2 is the variance of the workload process) respectively in relations (1) and (2), it was proved (Duffield, 1994) that the pair $\left\{\frac{W_t}{t}, \frac{t^2}{\sigma_t^2}\right\}$ satisfies a LDP with good rate function given by the *Legendre-Fenchel* transform

$$I(x) = \sup_\theta\{\theta x - \lambda(\theta)\}$$

of the *scaled cumulant generating function*:

$$\lambda(\theta) = \lim_{t\to\infty} \frac{\sigma_t^2}{t^2} \log E e^{\frac{t^2}{\sigma_t^2}\theta\frac{W_t}{t}}.$$

Moreover in (Duffield, 1994) the following fundamental asymptotic property is proved:

$$\lim_{b\to\infty} b^{-2(1-H)} \log P(Q > b) = -\delta,$$

where δ is given by:

$$\delta = \frac{1}{2\sigma^2}\left(\frac{C-m}{H}\right)^{2H} \frac{1}{(1-H)^{2(1-H)}}$$

Then we can say that, for large b values, the complementary probability presents a weibullian behaviour:

$$\log P(Q > b) \approx -\delta b^{2-2H}$$

The above relation, named *Norros's Bound* (Norros, 1995), represents a conservative approximation for the overflow probability when we deal with large values of buffer size, b. Figures 4 and 5 show that this bound is quite conservative and so we have to introduce a suitable additive constant which allows to achieve a tighter estimation.

We set this constant value equal to the probability of having a non empty buffer queue:

$$-\mu = \log P(Q > 0).$$

Then the estimation of the overflow probability with a buffer size equal to b is given by:

$$\log P(Q > b) \approx -\delta b^{2-2H} - \mu . \quad (3)$$

Once obtained the analytical estimation of the overflow probability, in the following section we validate it by means of discrete event simulations.

4 SIMULATION RESULTS

We arranged a set of simulations to verify the effectiveness of the modified bound (3) for different traffic loads and burstiness levels (i.e. H values). To this aim we used Random Midpoint Displacement algorithm (Lau, 1995) to generate several data sets representing sample paths of fractional Gaussian noise with different H values. We show the simulated queueing behaviour for two values of the normalised offered load which are very close to the desired working conditions of the network. For each simulation we assumed a mean arrival rate equal to 300 cells per Time unit and a standard deviation of 100.

Figures 4-7 show the behaviour of the complementary probability $P(Q > b)$ in logarithmic scale against b (so called *watermark plot*): we represent simulation results with dots and error bars (95% confidence interval) and the two analytical approximations with dashed lines (theoretical expression (3) and Norros's bound).

In each case we can see that, for values of b large enough, bound (3) is conservative and represents a tighter approximation for overflow probability than the well-known Norros's bound for different H and normalised load values. When low buffer sizes are considered, the asymptotic nature of bound (3) comes out; the simulation results show that the gap between the proposed relation and the simulated results grows as H increases.

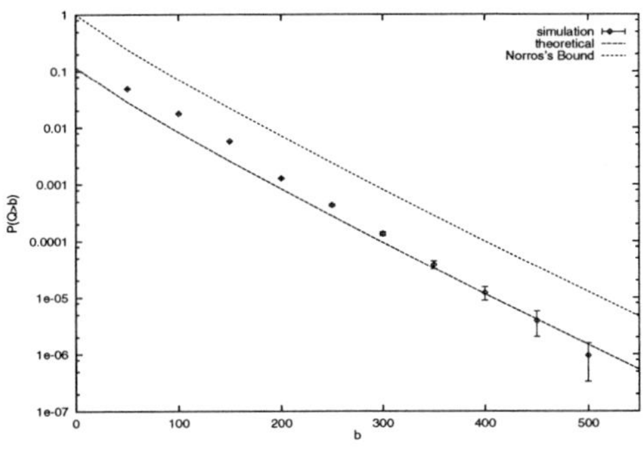

Figure 5. Complementary probability for H=0.55 and offered load 0.7.

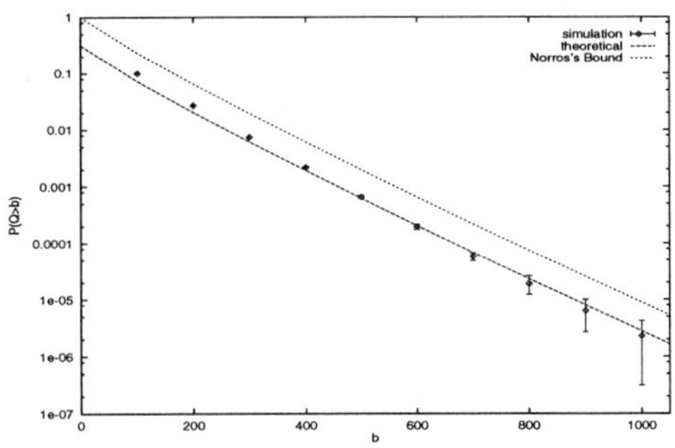

Figure 6. Complementary probability for H=0.55 and offered load 0.8.

Figure 7 highlights the impact of LRD on queueing performances. Moreover, by comparing the results with those shown in figures 4 and 6, we can see that an overflow probability around 10^{-3} is achieved with b=150, b=250, b=500 and b=3000 cells respectively for H=0.55, 0.65, 0.75 and 0.85 with the same offered load equal to 0.7. Thus the higher is H (i. e. the LRD) the more critical is the queueing behaviour.

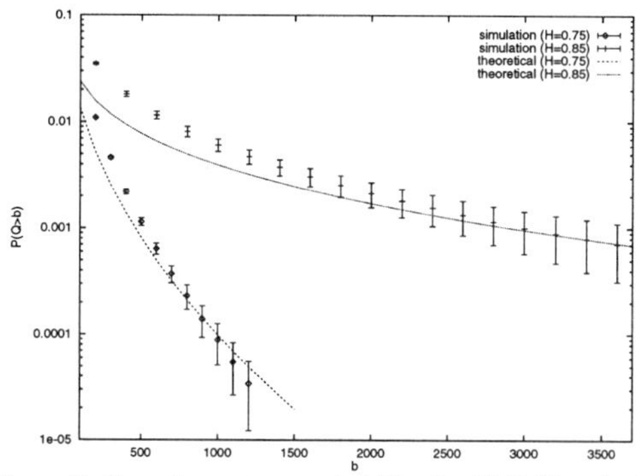

Figure 7. Complementary probability for H=0.65 and offered load 0.7 and 0.8.

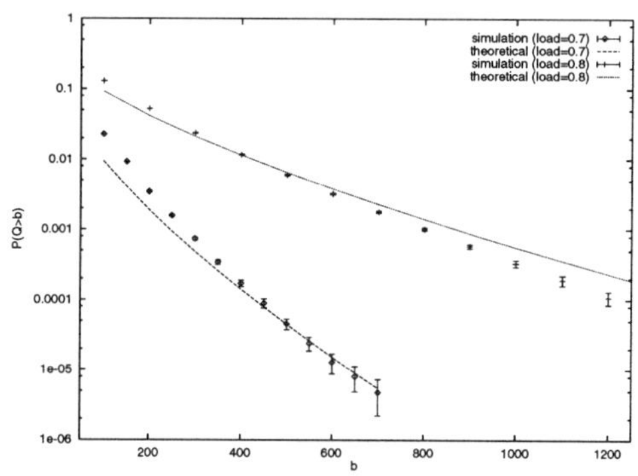

Figure 8. Complementary probability for H=0.75, H=0.85 and offered load 0.7.

5 CONCLUSIONS

In this paper we investigated the performances of an infinite buffer single server queue, with deterministic service rate, loaded by fractional Gaussian noise, a well-known LRD process. In particular we have tested the analytical estimation of the complementary probability determined using LDT. The main contribution of this paper is the introduction of a proper biasing in the results presented in (Duffield, 1994) that permits us to give a tight approximation for buffer overflow probability. This estimator could be useful in the context of preventive resource allocation techniques with particular interest for *call admission control* with stringent QoS requirements.

REFERENCES

Bucklew J. A. (1990), Large Deviations Techinques in Decision, Simulation, and Estimation, John Wiley & Sons, New York.

Dembo A., and Zeitouni O. (1993), Large Deviations Techniques and Applications, Jones and Bartlett Boston, London.

Duffield N. G., and Connell N. O. (1994), Large deviations and overflow probabilities for the general single-server queue, with applications, DIAS Dublin.

Giordano S., Pagano M., Russo F., and Tartarelli S. (1998), Efficiency of Importance Sampling Technique for Queueing Performance Evaluation with LRD Traffic, to be published in AEÜ Int. J. Electron. Commun.

Lau W. C., Erramilli A., Wang J. L., and Willinger W. (1995), Self-similar traffic generation: the random midpoint displacement algorithm and its properties, in Proceedings of the ICC '95, pp. 466-472, Seattle, WA.

Norros I. (1995), On the Use of Fractional Brownian Motion in The Theory of Connenctionless Networks, IEEE-JSAC, Vol. 13, N. 6, pp. 953-962.

Shwartz A., and Weiss A. (1995), Large Deviations for Performance Analysis. Queues, Communications and Computing, Chapman & Hall, London.

Willinger W., Taqqu M. S., and Erramilli A. (1996), A bibliographical guide to self-similar traffic and performance modeling for modern high speed networks, in *Stochastic Networks : Theory and Applications in Telecommunication Networks* (F. P. Kelly, S. Zachary and I. Ziedins, eds.), Vol. 4 of Royal Statistical Society Lecture Notes Series, pp. 91-104, Oxford University Press, Oxford.

Rosario Giuseppe Garroppo was born in Trebisacce, Italy 1969. In 1995 he received the Italian laurea degree in Telecommunications Engineering from the University of Pisa, Italy. Currently he is a PhD student at the Department of Information Engineering of the University of Pisa and his research activity deals with realistic traffic modelling and queueing performances evaluation.

Stefano Giordano was born in Pisa, Italy in 1963. He received the Italian laurea degree in Electronics Engineering from the University of Pisa in 1990 and the Dottorato di Ricerca (PhD) degree in Telecommunication Engineering in 1995.
Since the end of 1993 he is assistant professor at the Department of Information Engineering of the University of Pisa (Telecommunication Networks Group).
His main research and professional areas of interest are Broadband Communications, Telecommunication Networks Analysis, Design and Simulation, Multimedia Communications and Source Coding.

Michele Pagano was born in Lerici, Italy 1968. In 1994 he received the Italian laurea degree in Electronical Engineering from the University of Pisa, Italy, and in the same year he began his PhD studies. Since 1997 he is assistant professor at the Department of Information Engineering of the University of Pisa. His research activity deals with self similar modelling and wavelet analysis of multimedia traffic in broadband networks.

Gregorio Procissi was born in Montecatini Terme, Italy 1972. In 1997 he received the Italian laurea degree in Telecommunications Engineering from the University of Pisa, Italy, and currently he is studying at BRIMS Institute in Bristol. His research activity deals with Large Deviation Theory and self similar processes.

Franco Russo was born in Nola, Italy 1937 and he received his Italian laurea degree in Electronic Engineering in 1962 from the University of Pisa. From 1963 to 1981 he was researcher in the Italian National Research Council (CNR) and since 1981 he is full professor in "Digital Signal Processing". He is member of the Italian National Research Council (CNR) Group in "Telecommunication and Information Theory" and of the executive board of National University Consortium for Telecommunications (CNIT). Currently his main research area covers digital signal processing and telecommunication networks.

PART SEVEN

Advanced topics in optical networks

ATM and Optics

P. Boffi, G. Maier, M. Martinelli[+] and A. Pattavina[+]

CoreCom
via Ampére, 30 - 20131 Milan, Italy
Tel.: +39-2-2369131, Fax: +39-2-23691322
email: maier@corecom.polimi.it

[+]also with
Department of Electronics and Information,
Politecnico di Milano
P.za Leonardo da Vinci, 32 - 20133 Milan, Italy

Abstract

The contribute reviews aspects of the interaction between the asynchronous transfer mode (ATM) technology and the optical technologies. ATM for broadband networks presents some issues which result appealing for an optical approach. Some of them consist in the possibility of implementing ultra-high velocity ATM protocol functionality, in the development of free-space packet switching networks, in the solution of the access contentions to a shared transmission resource by deploying the wavelength resource. On the other hand, the development of the optical layer gives the transmission bandwidth availability which forecasts a larger penetration of ATM even at the network edge.

In the communication some examples of the ATM/optics interaction will be given. In particular it will be shown how to implement optically two basic functions of the ATM protocol, that is the Header Error Control and the Traffic Policing. Also, the interaction between a remote ATM network and the access network will be object of the presentation.

Keywords

All-optical switching, photonic packet-switching, ATM switching, header processing, node interface.

1 INTRODUCTION

In the last decades one of the most relevant phenomena in telecommunications has been the 'photonization' of ever increasing portions of the fixed communication infrastructure of the worldwide networks (Sato, 1996). This process began with the rapid deployment of optical fibers in almost all network levels as an alternative to copper wires. Technical and economical advantages of the new transmission medium were so strong that not only copper was almost completely abandoned for the new installations, but also a new branch of research in optics applied to telecommunications was fostered.

With the introduction of the optical amplifiers the optical fiber transmission systems evolved so rapidly that they are now a fully established technology, though still many issues remain to investigate. New records of transmission performance, such as nonregenerated distance, bit-rate and number of wavelengths, are continuously achieved by advanced systems, while commercial systems are always evolving toward high reliability and low cost targets.

In the present situation the 'photonization' process is on its way to enter a second phase, which is the penetration of optics into switching. If information propagates through the links of a network in an optical format, it seems quite convenient that it should not be converted to an electronic format in each node, to be converted in optics a second time on the next link: this main reason justifies the idea of optical switching. If this is true in principle, the realization of switching by means of optical technology has proved to be a nontrivial task, and in some cases quite difficult.

In the following of this contribution, after a short look at the perspectives of optical networks in general, we review the current status of research on ATM photonic switching. This particular aspect of optical networks can be considered nowadays an advanced topic compared to other solutions. In fact, even if many research activities have been dedicated to it, various critical issues remain unsolved. We will therefore review the main projects regarding all-optical ATM switching, mainly focusing on the node architectures, propose some new investigation objectives and outline what are in our opinion the 'hottest topics' for the future.

2 A PERSPECTIVE ON OPTICAL NETWORKS

The main feature which characterizes all-optical networks is transparency (Iannone, 1996): this property implies that optical signals can traverse more nodes and links without undergoing opto-electronic conversions (other properties such as scalability and modularity are also very important).

There can be different definitions or different levels of transparency. Transparency in the widest sense (*full optical transparency*) is attained when all network functions are independent of the signal modulation format and speed. In this case both digital and analog optical signal (as a subcarrier multiplex of CATV channels) can be switched by the nodes of the network. A more limited level (*bit-rate optical transparency*) concerns networks where only digital and intensity modulated optical signals can be routed, but processing performed by the nodes is independent of their bit-rate. When optical networks based on packet switching are considered, then another meaning of transparency can be given: *payload optical transparency*. In order to perform packet switching nodes must process only the header of the packets. Optical packet switching nodes can be therefore designed in such a way that they are transparent to the bit-rate of the information carried in the payload, as this part of the packet it does not have to be read or modified.

Research in optical switching led to the proposal of different solutions which exploit various levels of optical transparency.

In the access area all-optical switching has been employed in the passive optical network (PON) networks. Time-division multiplexed (TDM) PONs (Figure 1a) are the most developed solution (Faulkner, 1997). The TDM multiple access protocol required to manage the upstream traffic is implemented by electronic equipment located in the network terminations. Therefore the network cannot be considered truly optically transparent. An alternative solution for FTTH is WDM PON (Frigo, 1996) (Figure 1b): each user is identified by a wavelength channel, and the remote node consists of a wavelength-routing passive device, such as the wavelength grating router (WGR). These networks can be made *fully optically transparent*, and solutions have been proposed in which both digital (for voice and data traffic) and analog channels are switched (Frigo, 1997). A critical issue in this networking approach is the scalability, as the number of users is limited by the number of wavelengths that can be routed.

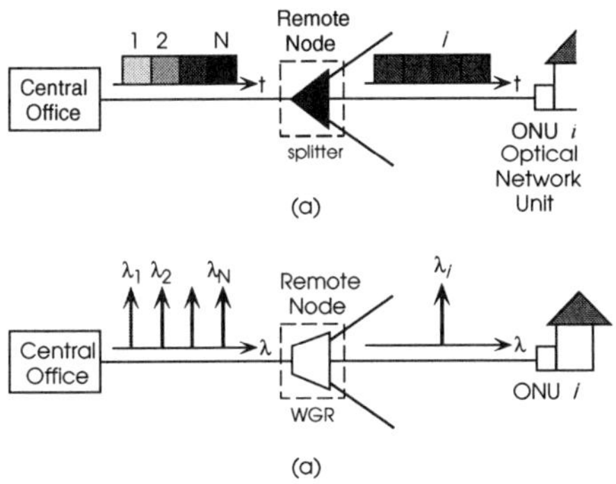

Figure I Passive Optical Network architecture: (a) TDM PON and (b) WDM PON.

The remote nodes of PONs perform only fixed routing, which is not a real switching function. All-optical switching performing a full switching functionality has been instead more developed in the transport area. Wavelength division multiplexing (WDM) technology is nowadays the main research topic in optical networks. A new network layer responsible of the switching of optical paths by means of optical cross-connect (OXC) nodes has been defined. In the WDM layer transparency is mainly intended as *bit-rate optical transparency*. The WDM core network is interfaced with higher network layers by means of electronic switches or SDH multiplexers (Sato, 1996). A scheme of the WDM transport layer is represented in Figure 2a. The optical paths in the transport network are circuit-switched. Besides routing also wavelength assignment must be performed by the network management system. Algorithms to evaluate optimal routing and wavelength assignment (RAW) have been analyzed in various studies (Banjeree, 1997), (Subramanian, 1997), also considering the case when OXCs are able to perform wavelength conversion of the optical paths. A general functional architecture of a WDM OXC is shown in Figure 2b. The node must provide switching connectivity in both space and wavelength domain. The reconfiguration time of optical paths is relatively long (in the order of milliseconds). In (Iannone, 1996) and (Sato, 1996) two main classes of OXC architectures have been identified, the first based on space switching and the second on wavelength switching. The first class is further subdivided into models based on a switching matrix and models relying onto the delivery and coupling concept. The blocking properties of these are also discussed there. The combined use of the space and wavelength multiplexing domains allows the exploitation of design criteria different from those typical of classical electronic switching matrices. The new wavelength multiplexing domain is also essential in photonic packet switching architectures, as we are going to explain later.

3 PHOTONIC ATM SWITCHING

Research in photonic packet-switching evolved in parallel with WDM. Many solutions have been proposed up to now, but none of them seems yet to have reached the same stage of development as WDM, for the reasons we will discuss in the following of the paper. Among all the solutions, we will consider in particular those relating to ATM switching techniques.
The investigation of ATM all-optical switching seems motivated by the opportunity of applying more general concepts of photonic packet-switching to a network model which is

well established and standardized. Moreover ATM seems to be a future-proof technology, as it allows transport features required by the future telecommunication services. Actually some disappointments about the initial expectations on ATM arise due to the fast diffusion of Internet and to a slower penetration of those services which rely upon severe quality of service targets and mostly justified the ATM features. Nevertheless, new architectural solutions such as IP-over-ATM (Decina, 1997) suggest not to abandon this subject.

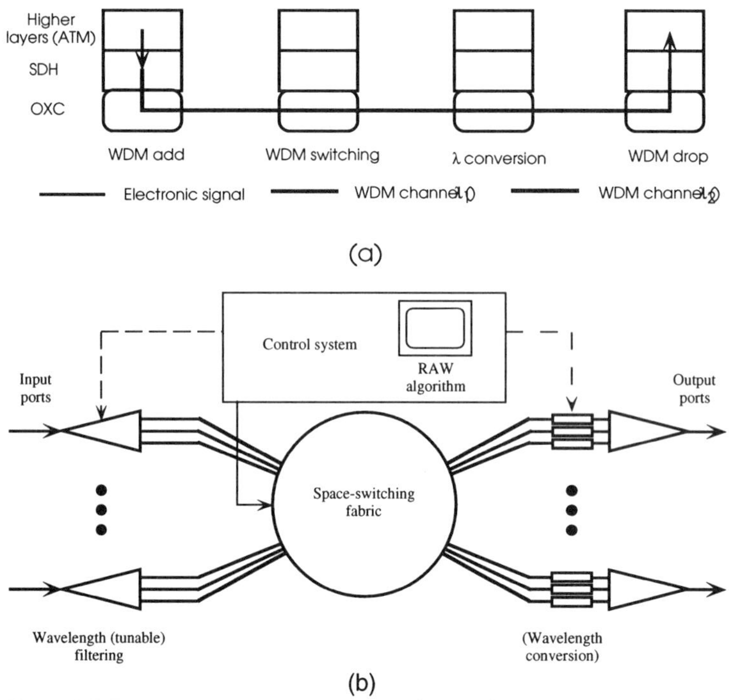

Figure 2 WDM transport networks: (a) a scheme of WDM optical path switching and (b) a functional scheme of an Optical Cross Connect.

An ATM network (de Prycker, 1995) is based on a paradigm quite different from the one we described for the WDM layer:

- Complete packet (or cell) switching operation has to be performed in this case in every node of the network (Figure 3a). This operation requires that the node identifies the cell, reads some routing information from the header and modifies it. Therefore *full optical transparency* can not be achieved in any way. The only transparency a photonic implementation would allow is what we called *payload optical transparency*.
- ATM is a connection-oriented packet switching technique. Virtual circuits and virtual paths are established, managed and terminated by the access nodes which perform more functions (belonging to the adaptation layer) than the internal nodes. The rate at which virtual circuits and virtual paths are set-up, routed and closed is dependent on the service characteristics and can potentially be of the order of the cell rate. This means that an ATM node not only must perform cell switching fast enough to adapt to the bit-rate, but also must be rapidly reconfigurable.
- The number of channels managed by an OXC is limited by the number of wavelengths per fiber. In ATM the number of virtual channel is by far much higher, since each channel is simply identified by a logical address carried by the VPI/VCI field of the header (2^{32} available values!).

In order to better understand photonic ATM switching a brief review of these basic functions according to ATM protocol could be helpful (Pattavina, 1998). The relevant layers of the protocol stack in this discussion are those whose functions are usually implemented by hardware, i.e. the physical layer and the ATM layer.

The ATM layer comprises the following functions:

- generic flow control
- cell header regeneration and extraction
- cell VPI/VCI translation
- cell multiplexing and demultiplexing
- policing and congestion control.

The physical layer is further subdivided in transmission convergence and physical medium dependent sublayers. The former comprises:

- cell rate decoupling
- header error control (HEC) sequence generation/verification
- cell delineation
- cell scrambling/descrambling
- transmission frame adaptation
- transmission frame generation/recovery.

The latter comprises:

- bit timing
- physical medium management.

A general architectural scheme of a node is represented in Figure 3b. The main difference between this and the other scheme reported for the WDM OXC in Figure 2b is the presence in this case of buffers and of processing interfaces.

The buffering capacity is essential for statistical cell multiplexing, and can be implemented in different ways and variously located inside the node architecture. The buffers and the space switching fabric together perform the core function of the switch, often referred to as multiplex switching.

All the other functions of ATM and physical layers are performed by the processing interfaces (one per each input/output port of the node). The most important of these functions is cell VPI/VCI translation: a peculiar characteristic of ATM is in fact the requirement of changing the logical address of a virtual channel or path at every node (label switching). HEC verification in particular is also very important since it is the basis also for cell delineation. Cell rate decoupling must be done when cells inside the multiplex switch assume a different format due to the addition, for example, of local routing information or guard-band intervals among consecutive cells. Transmission frame adaptation, generation and recovery are not necessary any more if we suppose that in the physical layer bit streams are organized directly as raw cells and are not carried by any transmission frame (like SDH). This simplification is quite important for an all-optical implementation, as we shall underline later on.

Summarizing, in an ATM switching node, beside multiplex switching, two other fundamental operations are performed by the processor interfaces: header reading and header writing. The payload of the cell is in no way involved in any of these ATM processing function, thus proving that *payload optical transparency* can be achieved by an optical implementation.

Most of research developed since today on photonic ATM switching has been dedicated to the multiplex switching. We will now briefly review the main solutions that have been proposed, trying to examine their architectural properties.

Two main approaches to photonic ATM switching fabric design have been followed. In the first approach only the switching fabric is considered, while buffering is left as a separate problem. In the second approach, instead, space switching and buffering are treated as an integrated problem.

Unbuffered architectures

In (Prucnal, 1993) simple 2x2 photonic switching elements (SEs) have been designed, which are capable of implementing cell self-routing on the basis of a local address and output port conflict resolution and have no buffering capacity. By employing these elements as building blocks, larger switching fabrics can be obtained. These fabrics can be defined partially connected networks, since a SE of a stage is not connected to all the SEs of the next stage. Thanks to the self-routing capability, different networks can be created with various blocking characteristic. The simplest is the Banyan network, which is blocking as is well known, since it provides a single path per inlet/outlet pair and each interstage link is shared by several inlet/outlet paths. To improve blocking performance deflection routing has been proposed: if two cells have to be routed onto the same interstage link of the fabric, one is actually routed towards the required destination, while the other is sent onto a sub-optimal path. Blocking can be removed if the connection requests to the switching fabric are previously sorted. The most known example of an architecture providing this feature is Batcher-Banyan, in which a Banyan network is put in cascade after a sorting network (Batcher). Photonic SEs designed for the Banyan network can be quite easily transformed into sorting elements, which are the building blocks of the Batcher sorter.

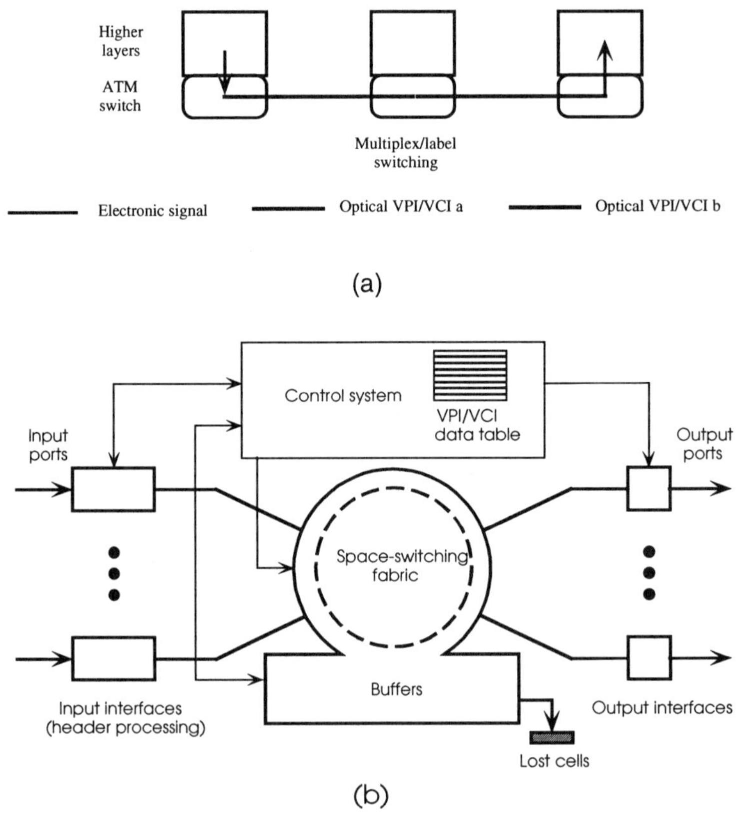

Figure 3 ATM networks: (a) a scheme of ATM switching and (b) a functional scheme of an ATM switching node.

Another type of unbuffered architecture can be obtained by connecting in various topologies output-multiplexing building blocks. These blocks have N inputs but $N \cdot K$ outputs: though they have no inner buffer, they resolve output conflicts, since up to K cells can be routed to the same output in the same time slot. In ATMOS an output-multiplexing architecture has been proposed, based on a building block called Multidimensional Switch (Eilenberger, 1995). In this block the output multiplex is easily realized by wavelength multiplexing more cells on an output fiber.

Buffered architectures

In the second approach to ATM switching design more complex architectures have been proposed. The concept common to all of them is that the complete switch should be obtained connecting stages of smallest modules, according to a fully connected topology (each module of a stage is connected to all the modules of the next stage), such as a Clos topology (Casoni, 1994). This design concept allows to distribute most of the control and management functions, thus simplifying the control system of the node. Each module is equipped with buffering capability, so that also buffering is distributed. In ATMOS a reference model of the overall switching system has been identified whose parameters are: 256 input/output ports with a bit-rate of 2.5 Gb/s per port (Masetti, 1996). The target performance proposed for the switch is: 0.5 Tb/s throughput with a load of 0.8 Erl per port. These dimensions and performance can be accomplished employing single modules which have around 16 input/output ports. The buffering capacity is then assigned to each module according to the cell-loss probability which must be achieved. These figures, evaluated in ATMOS project, can give the order of magnitude of the sizes also of other photonic ATM switching systems.

The switching module architectures can be classified according to the position of the buffers as: input, output and shared buffer modules.

Input buffer modules have been seldom developed (Varvarigos, 1997), since this structure suffers from the head-of-the-line blocking phenomenon.

Most of the proposals concern the architectural solution with output buffering. In these systems storage elements are usually realized by means of fiber delay lines. However different peculiar properties of optical signals are exploited in order to let all the output buffers be accessed by cells coming from all the input ports at the same time (speed-up).

An architecture proposed by NTT (Tsukada, 1996) is called Ultrafast Photonic ATM Switch (ULPHA). This switching module employs the time resource to realize the speed-up. This is achieved by optically compressing the cells in time in such a way that, if N is the number of inputs and T (s) is the cell duration, inside the node each cell has a duration T/N (s). Actually the system is designed to operate with electronic signals at the input ports, but in principle it could also operate on an all-optical flow, provided that the input electro-optical modulators are substituted with all-optical modulators.

Architectures proposed in ATMOS and KEOPS projects rely instead on wavelength to obtain the speed-up necessary to access the output buffers. The first module proposed is the Fiber Delay Line Switch (Gabriagues, 1995). In this architecture cells at the inputs are wavelength converted and each one is "colored" according to the destination output port, which on the other hand will have an optical filter to identify the correct cells. Then all the cells are wavelength division multiplexed in a common buffer structure of fiber delay lines. The position of each cell in the buffer is determined by an optical structure based on the broadcast and select concept. A second module has been proposed (Broadcast and Select Switch) (Masetti, 1996) not very different from the Fiber Delay Line, but which allows cell multicasting .

Another output buffer architecture proposed by NTT is Frontiernet (Sasayama, 1997). In this case variable wavelength converters at the input and tunable filters at outputs are employed. Moreover a WGR instead of a star coupler is used to multiplex signals. The functionality of this architecture is similar to the Fiber Delay Line, but broadcast and select structure is not required.

The third class of switching modules comprises shared buffer architectures. An example is the Fiber Loop Memory Switch (Cinato, 1995), proposed in ATMOS. The buffering structure is composed of a single recirculating fiber loop, which is shared by wavelength division multiplexing. A number of logical FIFO queues, one per output port, are created and managed by the (electronic) control system over the same physical loop memory. An input cell is converted to a wavelength assigned according to the position it must occupy in the logical FIFO queue corresponding to its destination. Tunable filters in the outputs select the correct cell in each time slot.

The implementation of the processing interface functionality by means of all-optical subsystems has been studied only in the last few years, mainly as a consequences of the advances in all-optical processing device technology. One of the first example of this new activity is the header regeneration and writing subsystem proposed by (Chiaroni, 1996). The subsystem is based on the semiconductor optical amplifier (SOA), which seems to be a promising device. Another example (which actually has been proposed a bit earlier) is the cell synchronization and alignment subsystem based on fiber variable delay lines (Prucnal, 1993). Even if the cell flow is optical, the control signals of the subsystem are still obtained by electronically reading the header of the cells.

4 OPTICAL IMPLEMENTATION OF ATM NODE SUBSYSTEMS

In this section we present some application areas regarding ATM node on which we focus our attention. We briefly propose different solutions where optics allows to achieve high performance operations on the signal. Our aim is just to demonstrate the feasibility and the advantages of employing optical technology in telecommunication switching nodes. Optics can offer several benefits over traditional electronic approaches based on serial processing of information, because of the opportunity to explore new and efficient architectures.

Novel architectures able to exploit in a very efficient way the capabilities of optics in order to perform ATM interface processing are described. The proposed examples constitute the premise of photonic subsystems building for communication ATM nodes.

VPI/VCI field recognition

In ATM nodes the channels are identified by a logical address carried by the VPI/VCI field of the cell header. In any hypothesis of self-routing packet switching it is fundamental to recognize optical information. Here we propose an optical technique able to perform the recognition of digital words in a passive way without any opto-electro and electro-optic conversion. The recognizer is based on an optical Vander Lugt correlator realized by means of volume holography (Gu, 1995). To use it the input data flow must be converted in a spatial word constituted by a 2-D pattern of optical bits.

In a Vander Lugt architecture realized by means of volume holography the optical signal which carries the information to be recognized is set at the front focal plane of a lens so that the signal Fourier transform is obtained at the rear focal plane, where a thick recording medium is centered. A coherent reference plane wave lights the holographic medium. The interference pattern induced by the signal and the reference beams is stored as a perturbation of the index of refraction. When a new signal is coming in input, it can be compared with the stored pattern by detecting the Fourier transform of the diffracted beam at the rear focal plane of a new lens. In this point we read the correlation between the input signal and the stored one. Large storage capacity together with real-time response makes volume holography a very promising candidate for next generation optical processing. Multiple information can be recorded within the same holographic medium using angle multiplexing and can be used like a database for signal recognition. The input signal is in fact compared simultaneously with all the stored data producing an optical output just in the spatial position corresponding to the recognized information (Figure 4).

In Figure 5 we present the experimental results in the case of recognition of the digital byte 191. The storage of 256 holograms in visible light corresponding to the whole set of bytes in a single $LiNbO_3$ crystal is performed. A complete recognition of a single digital word in real time is obtained using phase-coded input signals.

Header Error Control

Parallelism represents one of the most advantageous features for future optical switching nodes. It matches naturally the use of spatial bandwidth: each signal of a parallel data stream is transported through a spatial channel by a light beam. Many spatial channels can propagate together in the same volume with no crosstalk. Parallelism is fully deployed by means of free-space architectures.

We present an example of free-space architecture employed in ATM processing: a parallel header error control (HEC) sequence decoder performing error detection on the incoming ATM cell headers. We do not consider the correction-mode HEC operation in this design.

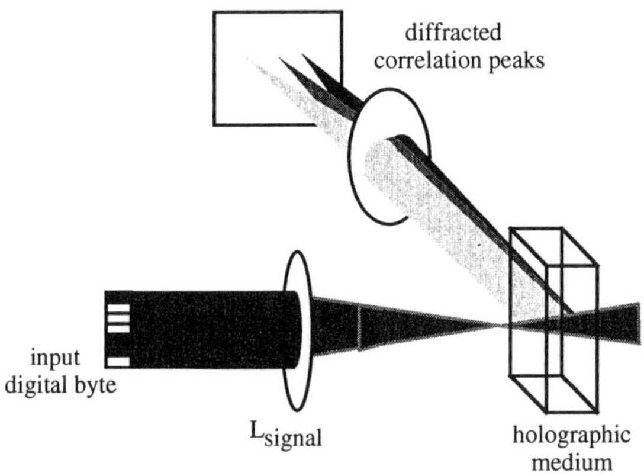

Figure 4 Volume holographic optical correlator useful for digital word recognition.

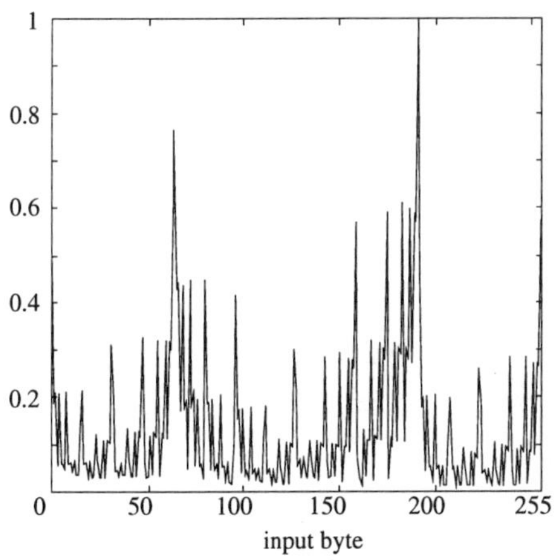

Figure 5 Correlation intensity related to the byte to recognize 191 as a function of the phase-coded input bytes. The correlation signals are detected in output of the volume holographic correlator and are normalized by the peak intensity of the autocorrelation signal.

The HEC is a function necessary to protect the information contained in the cell header from transmission errors. According to ITU-T standards (ITU-T I.423) the code in HEC processing is a "shortened" cyclic code. Using the classic serial algorithm (Benedetto, 1987), in which binary polynomials represent the header digital words, the decoding operation is carried out by computing a polynomial named syndrome, constituted by the remainder of the division between the received header to process and the generator polynomial. The serial decoding operation is generally implemented by electronic hardware, where a shift register performs the binary polynomial division operating at the bit-rate speed.

By resorting to matrix algebra we can find an alternative parallel decoding algorithm more suitable to exploit all the advantages of a free-space structure. In this formalism the error detection is achieved simply through the analysis of the syndrome vector **s**: if any bit of the vector is "1", then an error is present. The vector **s** is obtained multiplying the received header **y** (including 40 bits) by the transposed matrix **H'**:

$$\mathbf{s} = \mathbf{y}\,\mathbf{H'}$$

H (8x40 in size), the parity check matrix, is calculated directly from the generator polynomial of the code. Hence we can perform the HEC decoding completely in a parallel way, just by means of a single operation, with no flip-flops or memory required. Using modulo-2 arithmetic the operation $\mathbf{s} = \mathbf{y}\,\mathbf{H'}$ is the result of a sequence of scalar sums and products. The syndrome **s** can be evaluated from a **C** matrix (8x40) obtained through the AND operation:

$$\mathbf{C}(i, j) = \mathbf{Y}(i, j) \text{ AND } \mathbf{H'}(i, j) \qquad \forall \begin{cases} 1 \le i \le 8 \\ 1 \le j \le 40 \end{cases}$$

where the matrix **Y** (8x40 in size) is built by placing 8 identical columns side by side, each column equal to **y'**, the transposed of the row vector **y** corresponding to the received word. The vector **s** becomes:

$$\mathbf{s}(i) = \bigoplus_{j=1}^{40} \mathbf{C}(i, j) \qquad \forall \; 1 \le i \le 8$$

We propose a free-space implementation of the parallel algorithm described above (Figure 6). The received header signal is converted in a spatial form and then divided in 8 replica in order to obtain the 2-D **Y** matrix image. A simple array of beam-splitters can perform this operation. Then the AND operation between **Y** and **H'** is implemented simply by means of a spatial filter. This mask interrupts the optical propagation in relation to the "0" positions in **H'** and is transparent in coincidence of "1s". Because all the elements of **H** are "a priori" known, this operation is performed in a completely passive way. To perform multiple XOR operation along the 8 columns of **C** logic devices must be employed. 2-input XOR gates can be used in different configurations. The analysis of the syndrome vector can be obtain performing an OR operation on its 8 elements.

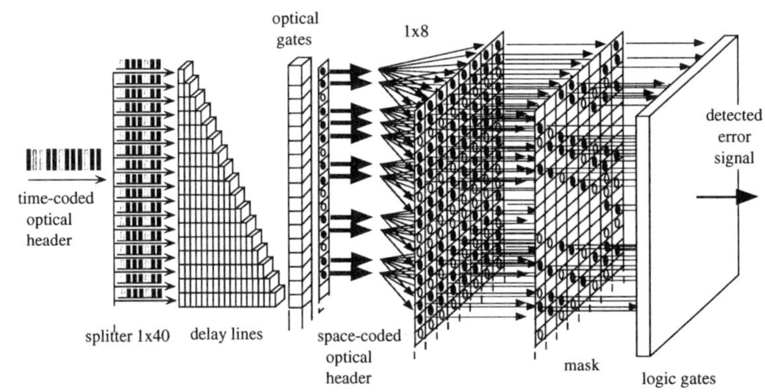

Figure 6 Scheme of the proposed implementation of the HEC optical subsystem.

The described structure is an example of a free-space architecture capable to perform a header processing function in an optical way. The utilized parallel decoding algorithm allows to eliminate the need of shift register and buffers. Parallelism allows the switching speed of each single logic gate operating in the subsystem to a certain data bit-rate is lower than the required one in the serial configuration. In fact in a system operating on a parallel data flow an equivalent bit-rate can be defined as the bit-rate of a single data channel of the flow multiplied by the degree of parallelism.

Policing function

In an ATM network the policing function controls the traffic of the transmitted data cells in order to protect the network against excessive congestion resulting in a degradation of the quality of the offered service. Policing allows the behavior of the traffic source to be restricted in accordance with the specifications negotiated in the contract between the users and the network provider.

Here we consider the policing function related to the control of the peak cell rate (PCR), i.e. the maximum rate at which the source can generate new cells. The virtual scheduling algorithm (VSA) is one of the algorithms standardized by ITU-T (ITU-T I.371). The traditional implementation of a policer based on VSA is realized by means of logic and mathematical elementary blocks performed by electronics. We explore new functional structures more suitable to exploit the capabilities of the optical devices.

In Figure 7 we propose a new scheme useful for the VSA optical implementation. The cells coming from the input link enter the policer. Here a cell input recognizer generates a recognition signal which acts on an optical controller. It must set the state ("open" or "closed") of an optical gate at the input of the system. According to the recognition signal the optical controller, by means of a suitable control signal, tailors a temporal window in which the connection input remains open. The temporal width of this window is established according to the maximum allowable cell rate negotiated with the network. If a new cell comes while the input gate is still closed, it is discarded.

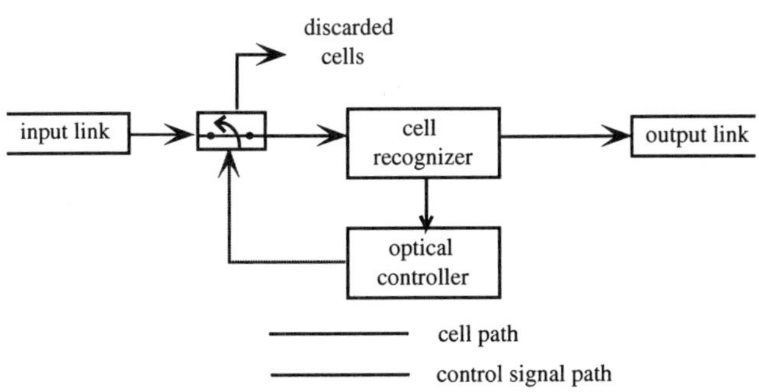

Figure 7 Logic scheme of the policer based on VSA.

The functional operators identified in the scheme described above are implemented using devices like couplers, delay lines and optical flip-flops. The cell-input recognizer can be realized in a simple way by means of an optical correlator capable to recognize a space-coded digital word. We have simulated the behavior of the system: Figure 8 and Figure 9 show the simulation results respectively in the case of a variable bit-rate input traffic and of a traffic with a bit-rate three times bigger than the allowed one. In our simulation the optical gate and the flip-flop in the controller are modeled like semiconductor optical amplifier (SOA) devices in interferometric structures (Chiaroni, 1995) in order to perform all-optical operations in which every control is optical.

5 CRITICAL ISSUES

The high complexity of a photonic ATM switching node that we tried to illustrate can be an obstacle for its future development, especially considering the competition versus WDM optical technology and also versus ATM electronic switching. Critical issues that arise involve technology, system and network aspects.

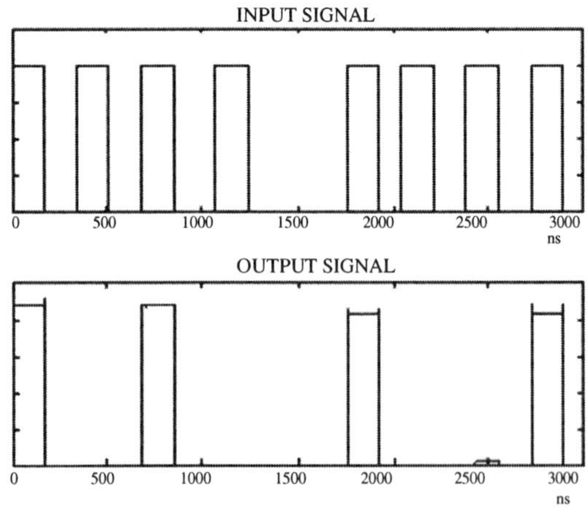

Figure 8 Simulation results of VSA policer operation. Upper trace: the input traffic at variable bit-rate. Lower trace: the restricted output traffic (maximum peak bit-rate STM-4).

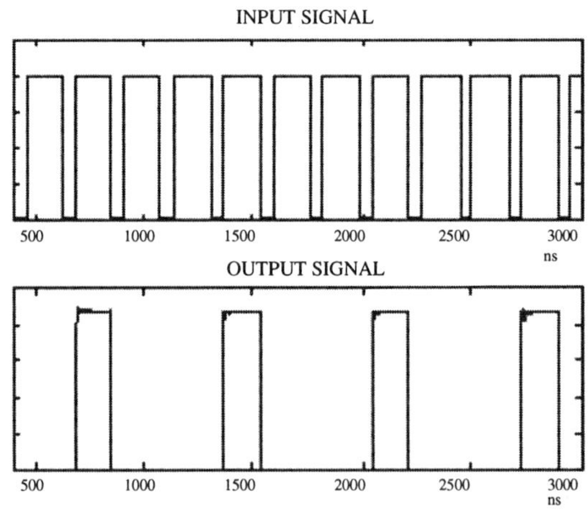

Figure 9 Simulation results of VSA policer operation. Upper trace: the input traffic at a bit-rate three times bigger than the allowed one. Lower trace: the restricted output traffic (maximum peak bit-rate STM-4).

If the advantages in terms of functionality of ATM versus WDM are clear, a debate issue is the advantage of an optical ATM solution versus an electronic implementation. It is true that the realization of a very large bandwidth node is a difficult task also in electronics (Butner, 1996), especially in those subsystems, like line interfaces, where bit-to-bit parallelization (the technique used to overcome the speed limitations of CMOS technology) can not be employed. However the costs of electronic components remain nowadays still much lower than that of optical devices. A final assessment of this point is still impossible, given the continuous evolution of both the technologies. Another penalty of optical technology is reliability. It has been demonstrated (Wosinska, 1997) that at present all the proposed ATM optical architectures have a reliability quite lower than that required by standards.

From a system point of view the advantages of all-optical ATM must be still investigated. Unfortunately the architectures described in the previous section has been tested only as prototypes with limited functionality: for instance the cascading of modules in multistage architectures has only been simulated, while all the experiments have been performed on a single module. Beside that, quite often only the multiplex switching fabric has been tested by experimentation. The all-optical subsystems proposed belonging to the processing interface, instead, have been tested separately or have not been tested yet. Often in experiments some all-optical subsystems are replaced by an electronic version, due to technical or economical reasons. If some functions of a photonic ATM node are implemented by electronics, the advantages of optics, such as *payload optical transparency*, may not be achievable. The fact that photonic nodes cannot still be tested as integrated complete systems prevents them from being experimented in a real network environment.

The positioning of photonic ATM in the network layer structure is not yet as clear as in the case of WDM. Various attempts, for example, have been made to define a frame structure which suites to optical packet switching better than the ATM cell, but which in the mean time is not too far from ATM standards. None of the proposals made gained since now an outstanding role.

Another problem still to solve is to better understand where in the network infrastructure all the advantages of photonic ATM could be better exploited.

In the transport area the interoperability with lower network transport layers should be taken into account. If ATM cells should be carried in SDH frames, for instance, the *payload optical transparency* offered by photonic ATM nodes would appear useless, since up to now SDH processing can only be done by electronic devices.

In the access area the conditions are quite different. We have studied the problem of introducing an all-optical ATM switch as the remote node in a fiber-to-the-home (FTTH) access network. Due to the simple tree topology of the FTTH an ATM switch located in the remote node behaves as a cell dispatcher for the downstream flow and as a statistical multiplexer for the upstream flow. While in the case of the cell dispatcher technology and system aspects are more interesting, the cell multiplexer behavior should be analyzed from a network performance point of view. In order do this we chose as a reference the TDM PON model defined in (Faulkner, 1997), which is also compliant to the standard draft (ITU-T 50-E). In this model an ATM upstream traffic is managed by means of a multiple access protocol called Global FIFO, implemented by electronic equipment located both in the central office and in the user network terminations (Optical Network Units, or ONUs). This protocol is based on dynamic time slot allocation on the basis of requests that each user periodically sends to the central office. The amount of bandwidth of the total upstream flow depends on the occupancy status of the local buffer in each ONU (user buffer). Cell collisions cannot occur, since time slots are properly allocated in a centralized way. In the case of the photonic ATM remote node there is no centralized remote assignment and cell collisions do happen. Therefore the node must be able to resolve cell collisions and must also be equipped with a certain buffering capacity in order to reduce the cell loss probability.

The study, conducted with various traffic source models (here only the constant bit-rate - CBR - case is shown, but results with other sources are quite consistent), has shown that the most relevant advantage of the ATM solution versus Global FIFO is in terms of delay (Figure 10). In fact in Global FIFO the TDMA protocol sets a minimum waiting time for each cell to be transmitted necessary for the permit-message exchange; the same constraint is not set by ATM. Also the delay variation is reduced in the ATM case (Figure 11). Such a neat advantage has not been found in terms of cell loss probability (Figure 12). The buffering capacity required inside the node is a trade-off between cell-loss probability and the node architectural complexity. Increasing the optical buffer in the node to reach low loss probability might lead to a cost increase which could be too high for such a cost sensitive area as the access network. Additional methods have also been investigated to allow a cooperation between the buffer in the ATM node and buffers located in the ONUs, which quite improves the situation.

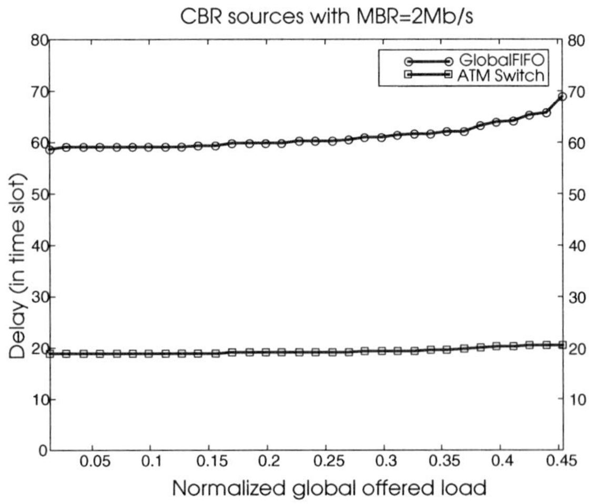

Figure 10 Comparison of cell mean delay between Global FIFO and ATM (CBR traffic).

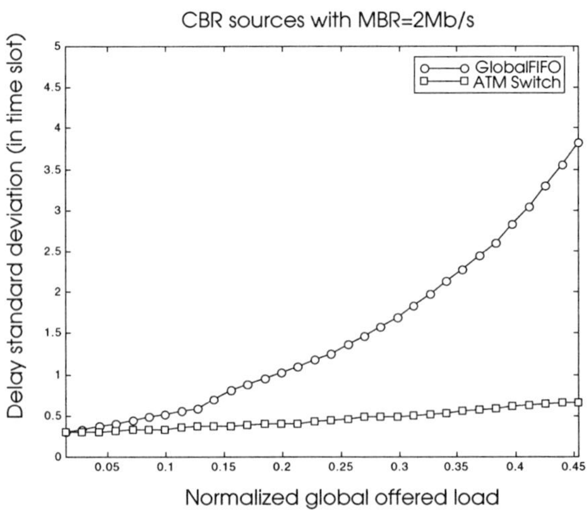

Figure 11 Comparison of cell delay standard deviation between Global FIFO and ATM (CBR traffic).

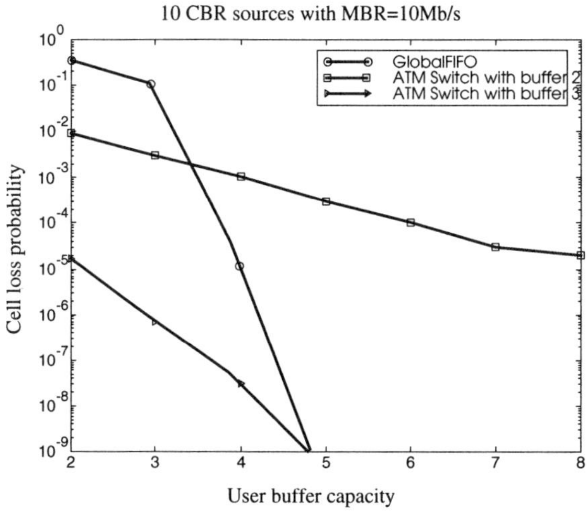

Figure 12 Comparison cell-loss probability between Global FIFO and ATM.

Acknowledgments

The authors wish to thank Professor Riccardo Melen of Politecnico di Milano for his support and Davide Piccinin, Lorenzo Ghioni, Giovanna Girola and Mauro Scappini for their contribution.

REFERENCES

Banjeree, D. and Mukherjee, B. (1997) Wavelength-routed optical networks: linear formulation, resource budgeting tradeoffs, and a reconfiguration study. *Proceedings of INFOCOM'97.*

Benedetto, S. et al. (1987) Digital Transmission Theory, Prentice Hall, Englewood Cliffs, N. J..

Butner, S.E. and Chivukula, R. (1996) On the limits of electronic ATM switching. *IEEE Network Magazine*, **10**, 26-31.

Casoni, M. (1994) Clos architectures for the design of large photonic ATM switches. *Proceedings of EFOC&N '94*, 106-110.

Chiaroni, D. et al. (1995) Sizeability analysis of a high speed photonic packet switch architecture. *Proceedings of ECOC'95*, 793-796.

Chiaroni, D. et al. (1996) Novel all-optical multifunctional regenerative interface for WDM packet-switching systems. *Proceedings of ECOC'96.*

Cinato, P. et al. (1995) Architectural analysis, feasibility study and first experimental results for a photonic ATM switching module to be employed in a large size switching network. *Proceedings of ISS '95*, 2 - C7.2.

de Prycker, M. (1995) Asynchronous Transfer Mode: Solution for Broadband ISDN. 3rd ed., Prentice Hall International.

Decina, M: and Trecordi, V. (1997) Convergence of telecommunications and computing to networking models for integrated services and applications. *Proceedings of the IEEE*, **85**, 1887-1914.

Eilenberger, G. et al. (1995) Concepts for an optical transport network and experimental results. *Proceedings of ISS '95*, C7.2.

Faulkner, D. et al. (1997) The full services access networks initiative. *IEEE Communication Magazine*, **35**, 58-68.

Frigo, N. (1996) Local access optical networks. *IEEE Network Magazine*, **10**, 32-6.

Frigo, N. et al. (1997) A WDM-PON architecture delivering point-to-point and multiple broadcast services using periodic properties of a WDM router. *Proceedings of OFC '97*, PD24.

Gabriagues, J. M. et al. (1995) Design and implementation of a gigabit ATM photonic switching matrix, *Journal of High Speed Networks*, **4**.

Gu, C. et al. (1995) Correlation patterns and cross-talk noise in volume holographic optical correlators. *Journal of Optical Society of America A*, **12**, 861-868.

Iannone, E. and Sabella, R. (1996) Optical path technologies: a comparison among different cross-connect architectures. *Journal of Lightwave Technology*, **14**, 2184–96.

ITU-T (1997) Temporary Document 50-E (PLEN), "G.PONB - Draft B", 7-18

ITU-T Recommendation I.371 (1993), "Integrated Services Digital Network (ISDN) - Traffic control and congestion control in B-ISDN"

ITU-T Recommendation I.423 (1993), "B-ISDN user-network interface - Physical layer specification".

Masetti, F. et al. (1996) Design and performance of a broadcast-and-select photonic packet switching architecture, *Proceedings of ECOC'96*, WeP.25.

Masetti, F. et al. (1996) High speed, high capacity ATM optical switches for future telecommunication transport networks. *Journal of Selected Areas in Communications*, **14**, 979-98.

Pattavina, A. (1998) Switching theory, architectures and performances in broadband ATM networks, John Wiley & Sons.

Prucnal, P.R. (1993) Optically processed self-routing, synchronization and contention resolution for 1-d and 2-d photonic switching architectures. *IEEE Journal of Quantum Electronics*, **29**, 600-612.

Sasayama, K. et al. (1997) FRONTIERNET: frequency-routing-type time-division interconnection network. *Journal of Lightwave Technology*, **15**, 417-29.

Sato, K.I. (1996) Advances in transport network technologies. Artech House.

Tsukada, M. and Nakano, H. (1996) Ultrafast photonic ATM (ULPHA) switch and a video distribution experiment, *Journal of Lightwave Technology*, **14**, 2154-2161.

Varvarigos, E.A. (1997) The "packing" and the "scheduling" packet switch architecture for almost-all optical lossless networks. *Proceedings of GLOOBECOM '97.*

Wosinska, L. and Thylen, L. (1997) Comparison of reliability performances of optical ATM-switches. *Proceedings of Photonics in Switching '97*, PD1-1.

BIOGRAPHY

Pierpaolo Boffi received his 'laurea' degree in Electronic Engineering from Politecnico di Milano (Italy) in 1991 and his Ph.D. degree in 1996 in Telecommunication Engineering from the same university. Since 1995 he has been a researcher at CoreCom. During 1997 he was visiting researcher at the Department of Electrical Engineering of the California Institute of Technology, Pasadena (CA). His current interests include optical signal processing and switching devices and their application to the development of communication network nodes.

Guido Maier received his 'laurea' degree in Electronic Engineering at Politecnico di Milano (Italy) in 1995 and is now attending the Ph.D. program in Telecommunication Engineering at the same university. He is also a researcher at CoreCom. His main area of interest are optical networks modeling and photonic ATM and WDM switching architectures.

Mario Martinelli was born in Mantova in 1952. He received the Laurea Degree in Nuclear-Electronics Engineering from Politecnico di Milano (Italy) in 1976. In 1978 he joined the Quantum Electronics Dept. of CISE Advanced Technologies, Milan (Italy). In 1991 he has been appointed Associate Professor of Optical Communications by Politecnico di Milano, where he activated the Italian first related course, and in 1993 founded the Photonic Lab at the Electronics and Information Dept. Since 1995 he is Director of CoreCom, a research consortium between Politecnico di Milano and Pirelli Cables, whose main area of activity id in the field of Optical Processing and Photonic Switching.

Achille Pattavina received the degree in Electronic Engineering (Dr. Eng. degree) from the University "La Sapienza" of Rome (Italy) in 1977. He was with the same University until 1991 when he moved to the Politecnico di Milano (Italy), where he is now Full Professor. His major research area has been the fast packet switching for broadband networks. He is author of the book "Switching Theory, Architectures and Performance in Broadband ATM Networks" (John Wiley & Sons). He has been Editor for Switching Architecture Performance of the IEEE Transactions on Communications since 1994. His current research interests are in the areas of optical networks and wireless networks.

Designing physical topologies for wavelength-routed all-optical networks

D. R. Hjelme[1,2] *and S. Eikedal Andersen*[1]

[1] *Department of Physical Electronics, Faculty of Electrical Engineering and Telecommunications, Norwegian University of Science and Technology*

[2] *Department of Photonics, SINTEF Electronics and Cybernetics*

N-7034 Trondheim, Norway

Tel: +47 73 59 27 22, Fax: +47 73 59 14 41

E-mail: Dag.R.Hjelme@ecy.sintef.no

Abstract

The influence of the physical topology design on the overall network cost is studied. Both optical cross connect cost and link cost, including cable and fiber/amplifiers, are considered. A number of regular and non-regular topologies with 9 to 24 nodes and node degrees between 2 and 6 are studies by computing the average number of hops, required to meet a uniform traffic demand, as function of node degrees and number of nodes. By relating the average number of hops to the most important cost factors, it is shown how the best topology (lowest cost) shifts from a sparsely connected topology if the cable cost dominates, to medium connected topology if the fiber cost is dominating, to densely connected topologies if the node cost is dominating. The lowest average number of hops, close to that of the ideal spanning tree, are found for highly non-regular topologies requiring long link lengths. If mapped on to a regular grid and using fiber by-passes at the node location such optimized topologies can be used to reduce the node complexity.

Keywords

Optical network, physical topology, wavelength division multiplexing

1 INTRODUCTION

Optimization and evaluation of the physical topology for wavelength routed optical networks (WRONs) is a difficult problem that is usually treated numerically. However, many of the network characteristics are to a large extent determined by a few average network parameters. For example, we have earlier shown (Hjelme 1996) that much insight can be gained using simple approximate analytical results relating a set of average network parameters. On reason that such simple analytical results work well, is that the design problem has a very large number of degrees of freedom. Often one finds that many apparently different designs performs very similarly. By reducing the number of degrees of freedom by limiting one self to global (or average) network parameters, we are able to focus on the dominating factors without being blinded by details.

Both the number of wavelengths and the number of hops required to satisfy the traffic demand are factors determining both feasibility and cost of the network design. The maximum values can be used to evaluate the feasibility, whereas the average values can be used to estimate the overall network implementation cost.

It has been shown that the physical connectivity of a network determines the wavelength requirements (Hjelme 1996, Baroni 1997); more densely connected networks require fewer wavelengths. In this paper we will focus on the average number of hops required, and show that this parameter determines many of the more important network characteristics. We will demonstrate that by knowing this parameters combined with a few simple analytical relationships, information on both transmission system requirements, node complexity, and network cost can be obtained.

2 SIGNIFICANCE OF THE AVERAGE NETWORK PARAMETERS

To be able to compare various topologies as candidates for WRONs, we have to consider both link dimensions, node complexity, and network cost.

2.1 Link dimensions

We consider a network with N nodes and N_L links. Each link consists of F_i (i=1 to N_L) fibers with W_{ij} (i=1 to N_L and j=1 to F_i) wavelengths. If we sum over all link capacities, we can express the total number of wavelength channels in the network as $\frac{1}{2}NDFW$, where D is the average node degree, F is the average number of fibers per link, and W is the average number of wavelengths per fiber. These wavelength channels are used to route the traffic demand T_i (i=1 to $N(N$-1)/2), in terms of wavelength channels, each requiring H_i hops, on all $N(N$-1)/2 bi-directional connections in the network. We define hops to be equal to the number of links traversed for one connection. If we sum over all connections, we can express the total number of wavelength channels needed to route the traffic as $\frac{1}{2}N(N-1)TH$, where TH is the average number of channels times the number of hops per connection.

The number of wavelength channels available in the network must be equal or larger than the number of channels required to route the traffic demand such that we can write the inequality

$$WDF/(N-1)T \geq H. \qquad (1)$$

We can use the equal sign only if there is no spare capacity. The amount of spare capacity required depends on the protection and restoration strategies used, a topic that is beyond the scope of this paper.

We have earlier shown that for a large class of topologies the ratio between the network diameter and the average link length, L, is given by (Hjelme 1996)

$$\text{network diameter} / L \propto H \qquad (2)$$

Our *first conclusion* is therefore that designing topologies that minimizes the average number of hops will reduce the transmission system requirement (with respect to number of wavelengths and fibers). We note that such designs will generally require longer link lengths than other topologies.

2.2 Node complexity

The left hand side of (1) is identical to the ratio between the total average node throughput traffic and the local add/drop traffic. Therefore, we can write

$$\text{throughput traffic} / \text{local add-drop traffic} \geq H \qquad (3)$$

Furthermore, using relationship (1), we can show (Hjelme 1997, Limal 1998) that the number of components in the optical cross connect (OXC) scales with H to the power one or two (depending on which components and OXC architecture (Jordan 1996) are considered), such that we can write

$$\text{OXC component count} \propto H^{\gamma}, \qquad \gamma = 1 \text{ or } 2 \qquad (4)$$

Our *second conclusion* is therefore that designing topologies that minimizes the average number of hops will i) reduce the required node throughput and ii) reduce the node complexity ($\propto$ number of components required in the OXC).

2.3 Network cost

We consider three contributions to the overall network cost i) fiber cost, ii) cable cost, and iii) OXC cost. The total link length scales with DL, and the total fiber length scales with $WDL \propto HL$. The OXC cost scales with the component number count, i.e. with H^{γ}. It then follows that *if* the overall cost of the network is dominated by

- the cost per fiber in the link, the overall network cost scales with HL,
- the cost per link (cable), the overall cost scales with DL, and
- the OXC cost the overall cost scales with H^{γ}.

The best topology for a given set of nodes and traffic demand therefore depends on the relative importance of these three factors.

Our *third conclusion* is therefore that that designing topologies that minimizes the average number of hops will help minimize the overall network cost.

3 THE RELATIONSHIP BETWEEN PHYSICAL TOPOLOGY AND AVERAGE NUMBER OF HOPS REQUIRED

H is determined by both the topology and the routing strategy used. In most cases, one will use a shortest path routing strategy. The choice of topology on the other hand is not obvious. The design space range from basic ring structures to fully meshed structures. For meshed topologies the choices include both node degree, i.e. the number of links to use, as well as the interconnect pattern, i.e. how we use the links to interconnect the nodes.

To learn more about the influence of the physical topology on the overall network cost we have computed the average number of hops required to route a uniform traffic demand (all-to-all connections) for a number of different physical topologies. The average number of hops was computed using Dijkstra's shortest path (in terms of hops) algorithm. We considered networks with number of nodes ranging from 9 to 24 with average node degrees ranging from 2 to 6. Both regular topologies and topologies designed to reduce the average number of hops were considered (minimum-*H* topologies). Our objective was to find out what can be gained by using the more complex minimum-*H* topologies compared to regular topologies.

The minimum-*H* topologies used here were designed with the objective to have an ideal spanning three from each node, i.e. a topology where the source node is one hop away from *D* nodes, which is one hop away from *D*-1 nodes and so on (Hluchyj 1991). We used an iterative procedure to design the topologies, but no true optimization was done. Still, the topologies showed an average number of hops very close to that of the ideal spanning tree. The deviation from this ideal situation is shown in Figure 1. For small networks (*N*<18), we have found topologies that give an *H* very close to the ideal. The deviations for *N*=24 is likely to be due to limitations in the search for the optimum topology.

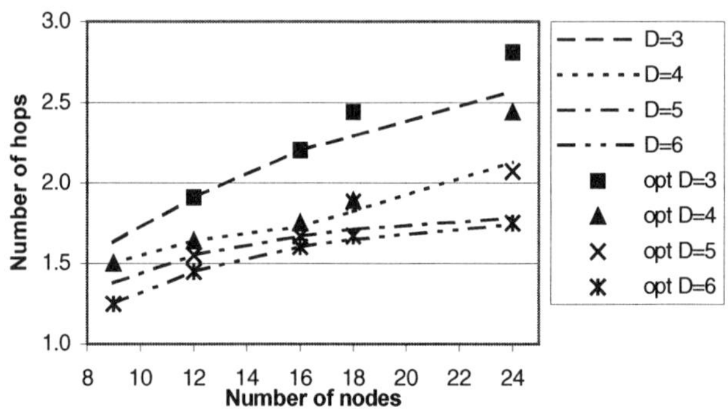

Figure 1: Comparison of the average number of hops for the 'optimized' topologies (markers) and for the ideal spanning tree (lines).

The results of the average number of hops simulations are shown in Figure 2 and 3. The different topologies considered (see legends used in Figure 2 and 3) included:

- *'opt D'* optimized to give a minimum average number of hops for a given node degree
- *'torus'* torus-like topologies where we have connected the network boundaries
- *'square'* regular grid topologies
- *'cylindrical'* regular grid topology on a cylindrical surface
- *'double ring'* two ring networks interconnected with 4 links
- *'triangular'* two ring networks interconnected with 3 links
- *'diamond'* regular grid limited to a diamond shaped area

For the first two topology categories all nodes have the same node degrees ('integer topologies'). The remaining topology categories have non-equal nodes and therefore non-integer average node degrees ('non-integer topologies').

From Figure 2 it follows that H increases sub-linearly with the number of nodes for all topologies. The larger the node degree, the lower the rate of increase. For clarity we have not included the results for D=2, i.e. ring networks, where H grows linearly with N. For a given network size, there is a significant reduction in H in going from D=2 to 3 and 4. However, increasing D to 5 and 6 results in only modest improvement (at least for the small to medium sized networks considered here). The optimized topologies reduce H by approximately 20% (D=3,4) and 10% (D=6).

Our <u>*fourth conclusion*</u> is therefore that for a given node degree we can reduce H significantly by optimizing the topology.

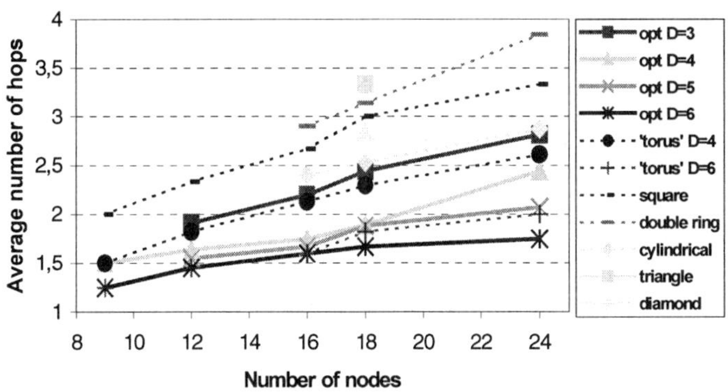

Figure 2: The average number of hops as function of number of nodes for different physical topologies. The various topologies are described in the text.

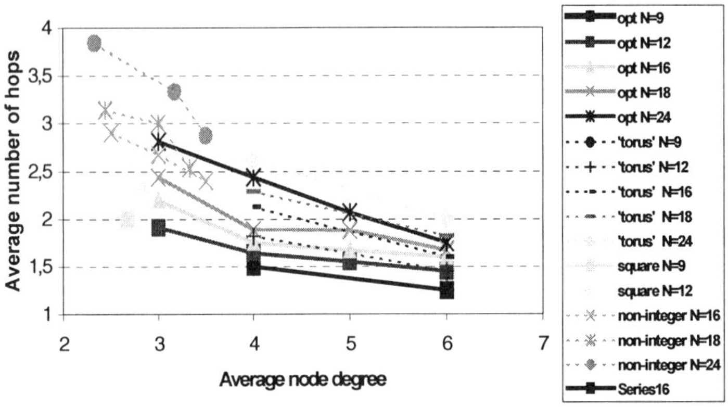

Figure 3: Average number of hops as function of average node degree for different physical topologies. For 'non-integer' topologies the data points are ordered such that the first point corresponds to 'double ring', the second to 'square', and the third to 'cylindrical'.

Figure 3 illustrates how H varies with the average node degree for various network sizes. For all networks considered here, H decrease with increasing node degree as expected. Again there is a large reduction in H going from D=3 to 4. The reduction is more modest for increasing D from 4 to 6. H decreases by approximately 15-25% for D increasing from 3 to 4 and from 4 to 6. The relative decrease of H is almost independent of network size and topology.

Our _fifth conclusion_ is therefore that for a given network size N, the most effective way to reduce H is to increase the node degree.

4 MINIMUM OVERALL NETWORK COST TOPOLOGY

The influence of physical topology on overall network cost is illustrated for an N=12 network in Figure 4. A ring, a regular grid, a torus, an optimized topology, and an optimized topology mapped onto a regular grid is compared. It follows that for this network size the best topology is a ring if cable cost dominates (DL), a square or torus if fiber cost dominates (HL), and an optimized topology if node cost dominates (H). The best topology therefore depends on the relative importance of these three factors.

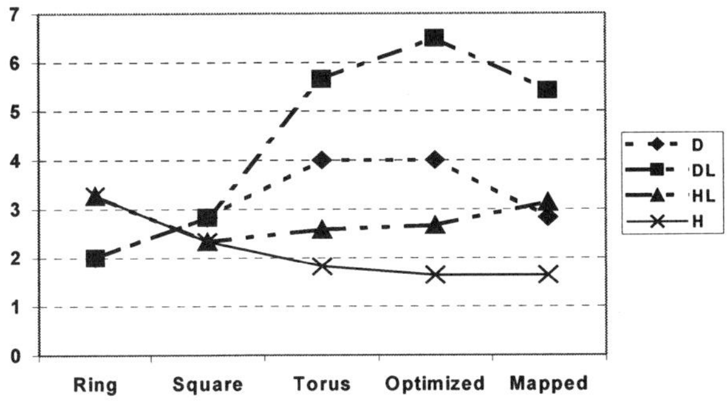

Figure 4: The three factors contributing to the network cost for an N=12 network. We estimate the link length by placing the nodes on a grid and computing the Euclidean length of each network link.

Minimum-*H* topologies are most effective at reducing the node complexity. They could therefore be attractive in situations where the main limitations are in the nodes. To simplify the network it is possible to map the minimum-*H* topology onto a regular topology. The minimum-*H* approach is then equivalent to regular topology approaches with fiber by-passes at the nodes. This will increase the fiber cost and reduce cable cost as shown in Figure 4.

5 CONCLUSIONS

We have shown how a small number of average network parameters can be used to guide the design and evaluation of WRONs. In particular, we have demonstrated that the average number of hops required to route the traffic demand is a key design parameter for such networks.

We have demonstrated that the overall network implementation cost is determined by the average number of hops required, *H*, the node degree, *D*, and the average link length, *L*. Using this, it is shown how the best topology (lowest cost) shifts from a sparsely connected topology if the cable cost dominates, to medium connected topology if the fiber cost is dominating, to densely connected topologies if the node cost is dominating.

The lowest *H* (and lowest node complexity) are found for highly non-regular topologies requiring long link lengths. These lowest *H* are found to be close to those of the case of an ideal spanning trees. It should be noted that the optimized topologies can be mapped onto regular grids by using fiber bypasses at the node locations. Compared to the ordinary regular grid networks, the optimized topologies will have reduced node complexities. This is achieved by properly sorting the traffic in the cable/fibers at the source node according to connections (destination node).

6 REFERENCES

D.R. Hjelme et.al. (1996), 'How many wavelengths does it take to build a wavelength routed optical network?', ECOC'97, paper MOA.4.2, Oslo, Norway, 1996.

S. Baroni et. al. (1997), 'Wavelength requirements in arbitrarily connected wavelength-routed optical networks,' Journal of Ligthwave Technol., vol 15, pp 242-251, 1997.

D.R. Hjelme, (1997) 'OXC components count as function of traffic scaling,' ACTS OPEN meeting, Oslo, Norway, 1997, unpublished.

E. Limal, K.E. Stubkjær, and D.R. Hjelme (1998), 'Optimization of the transmission system for wavelength routed optical network: trade-off between node complexity and transmission channel utilization,' submitted to NOC'98, Manchester, UK, 1998

A. Jourdan et.al. (1996), 'Design and implementation of a fully reconfigurable all-optical crossconnect for high capacity multiwavelength transport networks,' Journal of Lightwave Technol., vol 14, pp1198-1206, 1996.

M.G. Hluchyj et. al. (1991), 'ShuffleNet: an application of generalized perfect shuffles to multihop lightwave networks,' Journal of Lightwave Technol., vol 9, pp 1386-1397, 1991.

7 BIOGRAPHY

Dag Roar Hjelme received his Diploma degree in Electrical Engineering from the Norwegian Institute of Technology in 1982 and his Ph.D. degree in Electrical Engineering from University of Colorado (USA) in 1988. From 1988 till 1992 he was a Post.Doc and staff researcher with the University of Colorado. Since 1993 he has been with SINTEF, Trondheim (Norway), where he is currently a Senior Scientist. In 1995 he was appointed Adjunct Professor of Optical Communications at the Norwegian University of Science and Technology.

Stig Eikedal Andersen received his Diploma degree in Electrical Engineering from the Department of Physical Electronics at the Norwegian University of Science and Technology, Trondheim, in 1997.

Digital chaos synchronization in optical networks

A. González-Marcos and J.A. Martín-Pereda

ETS Ingenieros de Telecomunicación,
Universidad Politécnica de Madrid
Ciudad Universitaria s/n. 28040 – Madrid, Spain
Telephone: +34.1.336.73.04 , Fax:+34.1.336.73.19
jamp@tfo.upm.es;agonmar@tfo.upm.es

Abstract

A new method to obtain digital chaos synchronization between two systems is reported. It is based on the use of Optically Programmable Logic Cells as chaos generators. When these cells are feedbacked, periodic and chaotic behaviours are obtained. They depend on the ratio between internal and external delay times. Chaos synchronization is obtained if a common driving signal feeds both systems. A control to impose the same boundary conditions to both systems is added to the emitter. New techniques to analyse digital chaos are presented. The main application of these structures is to obtain secure communications in optical networks.

Keywords

Chaos syncronization, secure communications, digital chaos.

1 INTRODUCTION

A considerable interest has appeared in the last years concerning the application of chaotic circuits in order to obtain secure communications. The broad-band nature of chaotic signals makes them tempting for use in this area and in spread-spectrum situations. The fact that such signals emanate from a deterministic dynamic system lends to the hope that one will also be able to control them sufficiently for many uses. Taken as isolated systems, chaotic dynamics appears to offer many impediments to anyone attempting to put them to use. The main problem to be addressed is then how to synchronize chaotic circuits. It is well known that two identical circuits are able to offer similar chaotic outputs. But if these circuits are not synchronized their output signals are not valid to be employed in a communication system. The main reason is the strong dependence of the obtained chaos on the initial and boundary conditions of the chaos generator circuit. Two chaotic signals with the same characteristics may have different values at any particular time if they are generated independently. Some additional conditions have to be imposed to the system in order to obtain identical chaotic signals at any time. One of these conditions is the synchronization.

Several attempts have been made in this direction. The idea that chaotic systems could synchronize was first put forth in a paper more than ten years ago (Afraimovich, 1986). Several authors have followed the lines indicated in that paper. Pecora (1990) and Carroll (1991) demonstrated the possibility of sinchronizing chaotic subsystems with a common driving signal. Their idea was to decompose the chaotic dynamical system in two subsystems, "driving" and response" subsystems. The driving subsystem is composed by two state variable components whereas the second one just has one and uses as input signal one of the state components of the first subsystem. Some examples following this idea and schemes, using Chua's circuits, are reported in the literature (Madan, 1993).

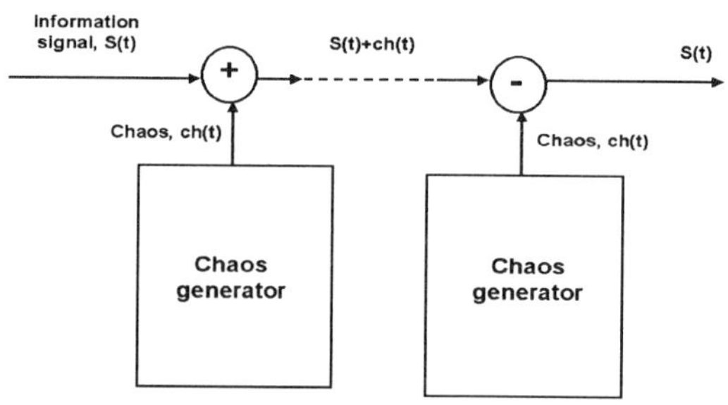

Figure 1 Basic configuration of the communication system

Another additional point needs to be considered. It concerns the characteristics of the chaotic signal to be employed. Almost in any of the reported situations the obtained chaos is analog. Although all physical systems are really analog, communication and computer systems are nowadays digital. The way to use analog signals in digital systems in to make the conversion analog/digital. This idea has been employed in any case where an application to communication is needed. It should be useful to obtain digital chaos from the very beginning of the process and to employ it with the same requirements of the information signals.

The purpose of this paper is to present a way to obtain digital chaos and how to synchronize two chaotic systems. The scheme of the proposed system is shown in Fig. 1. Two identical chaos generators, A and B, are located at emitter and receiver. Information signal is added at the receiver and transmitted to the network. This composed signal is detected at the receiver and processed with another chaotic signal obtained there. The resulting signal is the information generated at the emitter. This configuration is the basis of our system.

2 OPTICAL PROGRAMMABLE LOGIC CELL (OPLC)

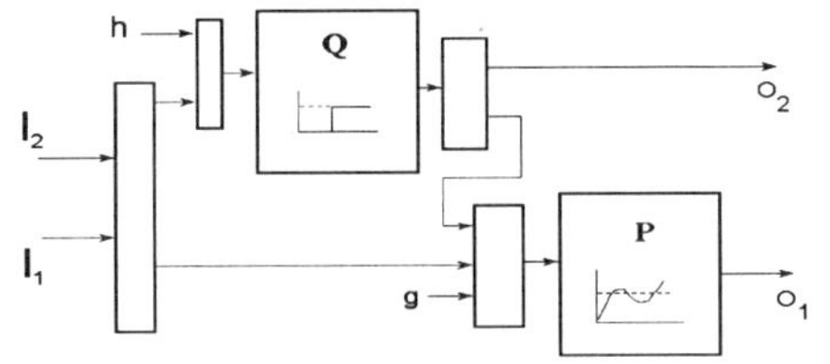

Figure 2 Internal configuration of an OPLC.

The main block of our chaos generator is an Optically Programmable Logic Cell employed previously by us as a part of a possible optical computer (Martín-Pereda, 1992). Although this structure has been reported in several places, some of its principal characteristics will be here presented again. Its main characteristic is the logic processing of two input binary signals, governed by two control signals. Two outputs give logical functions of these inputs. The type of processing is related to the eight main Boolean Functions, namely, AND, OR, XOR, NAND, NOR, XNOR, ON and OFF. The programmable ability of the two outputs, as it has been described, allows the generation of several data-coding for optical transmission. Moreover, as it was shown, this circuit has the possibility to the generation of periodic and even chaotic solutions. A precise analysis of the output characteristic versus the main variable parameters, as control signal level and data signal level, has been reported (González-Marcos, 1995).

With this configuration, the above mentioned digital character of the signal is directly obtained. Its main blocks are shown in Fig. 2. Two devices with a non-linear behaviour, P and Q, compose the circuit. The outputs of each one of them correspond to the two final outputs, O_1 and O_2, of the cell. Four are the possible inputs to the circuit. Two of them are for input data, $\mathbf{I}_1$ e $\mathbf{I}_2$, and the other two, $\boldsymbol{g}$ and $\boldsymbol{h}$, for control signals. The way these four inputs are arranged inside the circuit, is also represented in Figure 2. A practical implementation we have carried out of the processing element has been based on an optoelectronical configuration. Lines in Fig. 2 represent optical multimode fibers. The indicated blocks, placed in order to combine the corresponding signals, are conventional optical couplers. In this way, optical inputs arriving to the individual devices are multilevel signals. The characteristics of the non-linear devices are also shown in Fig. 2. Device Q, corresponds to a thresholding or switching device, and device P is a multistate device, being the response of this non-linear optical device the one represented in Fig. 2. This response is similar to the behaviour of a SEED device (Miller, 1985).

3 CHAOS GENERATION FROM AN OPLC

A non-linear behaviour is expected if some kind of feedback is applied to this cell (Ikeda, 1979), (Okada, 1981), (Neyer, 1982). The feedback we have applied to the system, among the different possibilities, is the one going from the output O_1 of P-device (see Fig. 3) to the control input, **g**, of P-device. No other additional control signal has been used. A chaotic output is obtained when the internal response time is made equal to zero or is much smaller than the external one. Some results have been reported by us (González-Marcos, 1996).

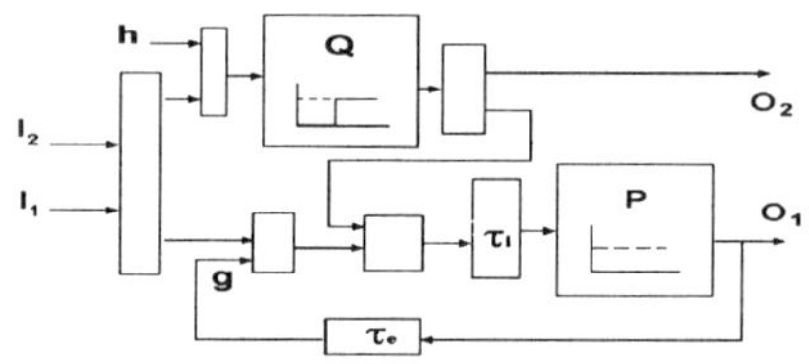

Figure 3 Optical Programmable Logic Cell with feedback. Chaos Generator.

In order to characterize the obtained chaotic signal, conventional methods are difficult to be applied here. This is due to its digital character (Martín-Pereda, 1995b). The results constitute a Time Series from where a chaos measure should be obtained. But the correct phase-space representation is not possible to be grasped from these results in a straightforward way. We do not even know what the adequate phase-space variables are and it is not even known how many variables are needed to fully describe the dynamics of this particular system. There is fortunately a partial answer to this problem that has been applied successfully in a large number of experimental investigations. The basic idea is that if the fundamental phase-space variables are **x** and **x'**, to study the evolution of the system numerically, x and x' have to be follow as functions of t. But since $x' = [x(t+\Delta t)-x(t)]/\Delta t$ in the limit as $\Delta t \to 0$, a knowledge of $x(t+\Delta t)$ is equivalent to a knowledge of x'. In other words, a knowledge of a trajectory of points $[x(t),x(t+\Delta t)]$ is equivalent to a knowledge of the trajectory of points $[x(t),x'(t)]$. As a consequence, a phase-space trajectory

$$x(t) = [x_1(t), x_2(t),..., x_n(t)] \qquad (1)$$

is replaced by a trajectory in an artificial phase space with points given by

$$y(t) = [y(t), y(t+\Delta t),..., y(t+m\Delta t)] \qquad (2)$$

where y(t) is any one of the phase-space variables $x_i(t)$. Thus from a set of measurements of a single quantity y(t) we can construct a sequence of points in an artificial phase space

$$x(t) = [y(t), y(t+\Delta t),..., y(t+m\Delta t)]$$
$$x(t+\Delta t) = \{y(t+\Delta t),..., y[t+(m+1)\Delta t]\} \quad (3)$$

With the data we have, the first problem to solve is how to operate with our digital signal where just two values, "0" and "1", are present. If we adopt just this output as possible values for **y**, the resulting plot at the phase space should be concentrated on just four points, namely, (0,0), (0,1), (1,0) and (1,1). Almost no information could be obtained from it. Hence a new technique has to be implemented.

The method we have adopted is to group sets of four consecutive bits and to convert them to their corresponding hexadecimal values. Hence, a sequence of zeroes and ones is converted to a new string of hexadecimal values, namely, 0, 1, 2, ..., 15. For example, "0010" would be a "2", "1001" a "9" and "1110" a "14". The total number of data is divided by four, but much more information can be obtained from them than with simple binary signals. A diagram, similar to the t_{i+1} versus t_i in analogue signals, can be drawn in this way. In the case of periodic signals, a closed configuration is obtained. But in the case of chaotic signals, no definite pattern would be obtained. Some results are reported in González-Marcos (1996).

A further point needs to be considered. It is the one concerning the justification that the preceding quantity, namely the hexadecimal sequence, represents the same behavior of the system than the previously obtained binary one. But this situation is equivalent to the reverse one: to convert a chaotic analog signal into a digital one. As it is well known from digital communications, any analog signal can be quantized and from this quantization to obtain a digital signal with the same main properties than the initial analog one. In our present case, we have accomplished the opposite operation, namely, to convert a digital signal into an analog quantized one with sixteen possible levels. And, according to digital communication signal processing theory, both representations are equivalent.

4 SYNCHRONIZATION OF CHAOTIC OPLCS

If two identical cells with feedback, as the above mentioned, are parallel connected (Fig. 4) and the same signal arrives to their inputs, an identical chaos is obtained at their outputs. This situation corresponds to two identical and ideal configurations working under identical conditions.

The behaviour becomes critical when the simulation tries to be close to a real situation. In this case, if both systems are not feed by exactly the same signal, the obtained outputs, although chaotic, are different. Hence, no possible relation between then should be feasible.

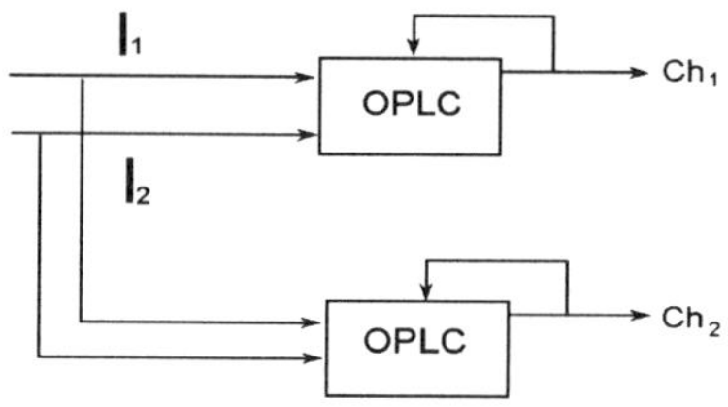

Figure 4 Basic blocks for chaos syncronization

In a general situation, both systems, emitter and receiver, are located at different places. As a consequence, there is no possibility to introduce exactly the same input signals to their corresponding input ports. This is because although a common signal generator could send the same train of pulses to both cells, the arriving times to them can be different. The time need to get the first cell is known if this generator is at the receiver place. But the time when the signal arrives to the second cell may not be known. This is the most general case. Several solutions could be implemented to overcome this fact. The solution we have adopted is presented in Fig. 5, and it was partially reported previously (Martín-Pereda, 1996).

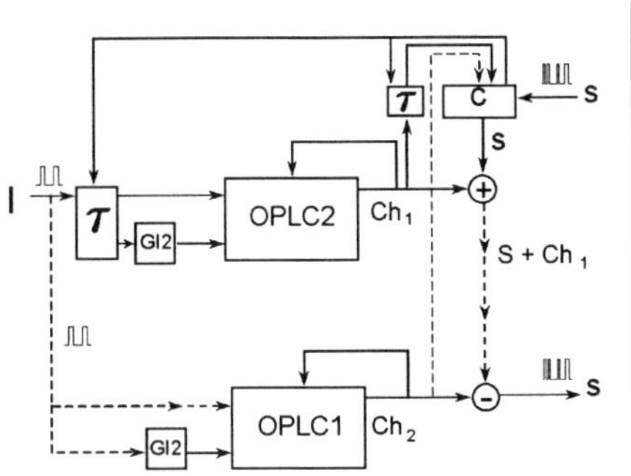

Figure 5 Schematic diagram of the communication system

The same driving signal is sent to both cells. A delay time, τ, is added before this signal gets into the emitter cell. This time has to be equal to the fly time from emitter to receiver. But this time is not known. Hence has to be changed until both times are the same. The way to change is imposed by the signal given by comparator C. There, the chaos signal coming from OPLC2 is compared with the chaotic signal obtained from OPLC1 and depending on the difference between these two signals, the order is given. As before, a delay time τ is added to the output from the OPLC1.

Two methods have been implemented in order to know the time delay magnitude to be added at the receiver. The first one is shown in Fig. 6.a. Chaos signal from OPLC1 is represented at the x axis and the one from OPLC2 at the y axis. A hexadecimal representation, as before, was taken. The second method is just to represent the difference "Chaos from OPLC1 – Chaos from OPLC2)" versus time. An example is given at Fig. 6.b.

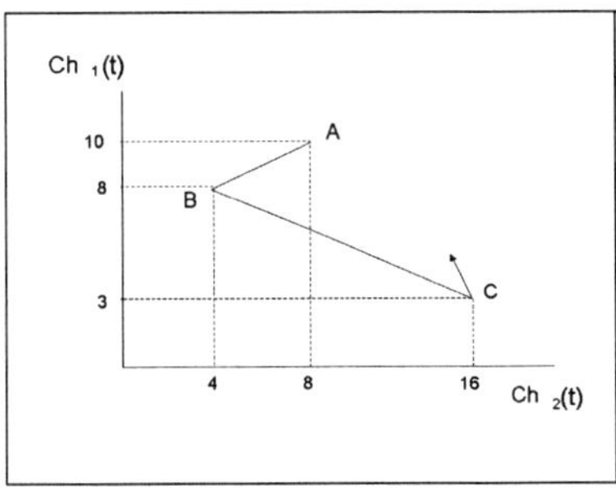

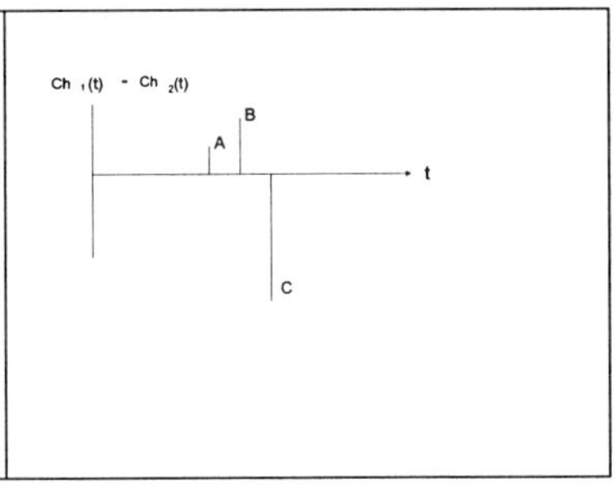

Figure 6.a

Figure 6.b

Another point needs to be considered. Because the strong dependence with respect to the initial conditions, it is necessary to reset to zero every time the delay time is changed. The variation should follow a line like the one represented in Fig. 7.

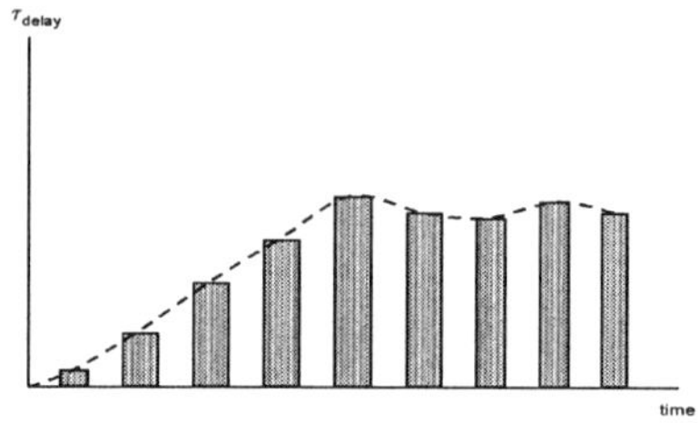

Figure 7 Delay time evolution to syncronize chaos generators en figure 5.

Some particular cases are represented in Fig. 8.a-d. A constant delay time equal to 20 time units has been taken in every case. The change between different representations is in the internal delay time added. An interesting fact needs to be pointed out. As it is seen, the main characteristics of the obtained graph, when delay time is smaller than fly time, is equal to the obtained when is larger by the same amount, but with a different angle with respect to the co-ordinate axis. This gives us an indication of the situation of the guessed delay time imposed to OPLC1. A similar result, but just with an indication of the chaos difference magnitude, is obtained from the above-mentioned second method.

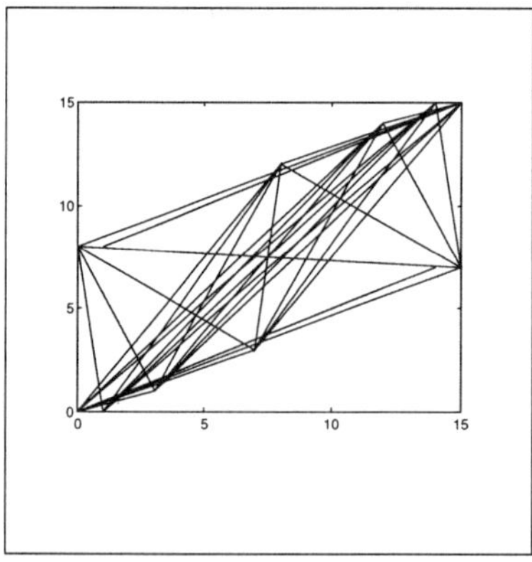

Figure 8.a Time delay = 19

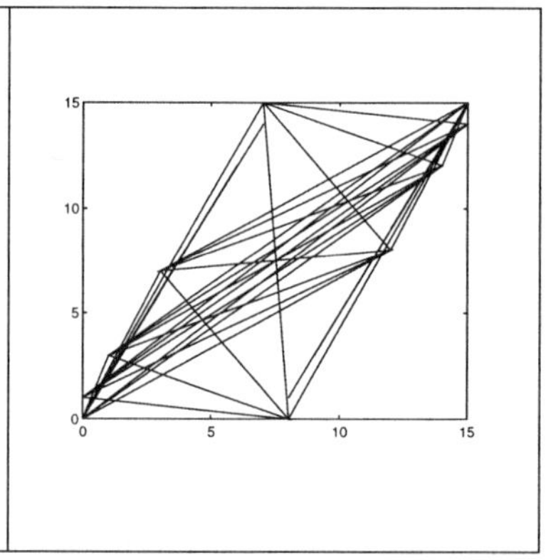

Figure 8.b Time delay = 21

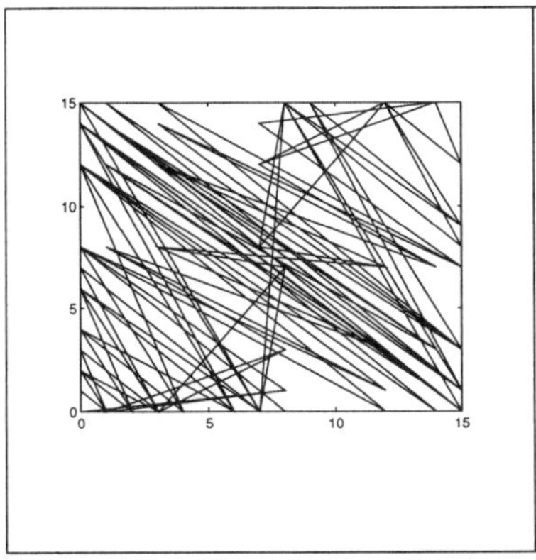

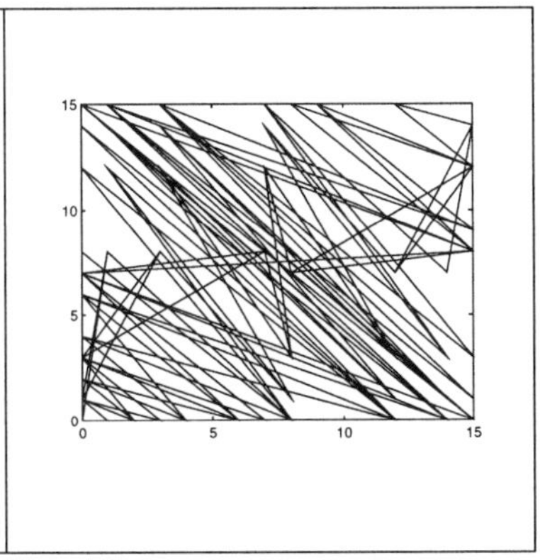

Figure 8.c Time delay = 25 **Figure 8.d** Time delay = 15

The final result, for a fly time of 20 time units and an initial imposed delay time of 15 is shown in Fig. 9. It shows "Chaos from OPLC1 minus chaos from OPLC2" versus time. As it can be seen, after a certain number of changes in delay time, synchronization is obtained.

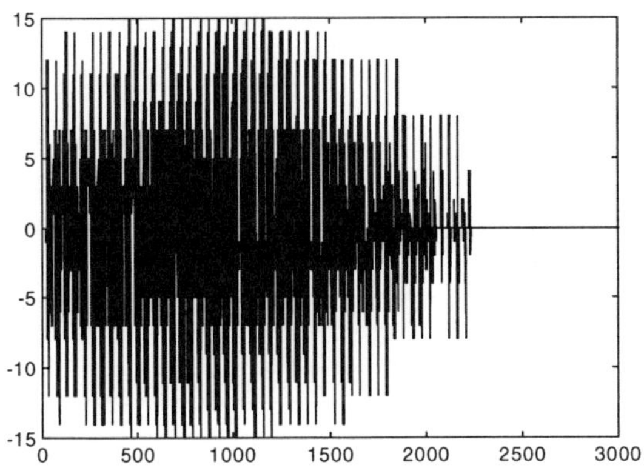

Figure 6 Substration of chaos signals from both channels.

5 CONCLUSIONS

A new method to synchronize chaotic circuits has been reported. The method is based on the use of Optically Programmable Logic Cells as chaos generators. Although these cells have been studied previously, some further studies are needed in order to determine the characteristics of the obtained chaos when the initial driving signal changes. The dependence with its parameters deserves also a deeper study. Moreover, the added delay time has been set to a certain value and if it is not the adequate one, the cell is reset to zero and the process start again. An study of how small and continuos changes affect the results are needed.

Finally, another point should be considered. It concerns the transfer of information among biological systems as well as their behaviour. Some pathologies and malfunctions in living bodies can be analyzed with similar methods as the above reported (Martín-Pereda, 1995 a). This area deserves a further study in order to approach these two fields, communication systems and communication between neurons.

Acknowledgement

This work has been partially supported by the Community of Madrid under research project 07T/0017/1997.

6 REFERENCES

Afraimovich, V.S.; Verichev N.N. and Rabinovich, M.I. (1986) Stochastic synchronization of oscillations in dissipative systems, Inv. VUZ. Rasiofiz. RPQAEC **29**, 795-803, 1986.

Carroll, T.L. and Pecora, L.M. (1991) Synchronizing Chaotic Circuits, *IEEE Trans. on Circuits and Systems,* **38**, 453.

González-Marcos, A. and Martín-Pereda, J.A. (1995) Sequences of bifurcations and transition to Chaos in an Optically-Processing Element, in *Optical Computing* (ed. B..S. Wherrett), . Inst. Phys. Conf. Series. No. 139 Part IV.

González-Marcos, A. and Martín-Pereda, J.A. (1996) Digital Chaotic Output from an Optically-Processing Element. *Optical Engineering*, **35**, 525-535.

Ikeda K. (1979) Multiple-valued stationary state and its instability of the transmitted light by a ring cavity system. *Optical Communication*, **30**, 257-261.

Madan (1993), in *Chua's Circuit: A Paradigm for Chaos* (ed. R.N. Madan) World Scientific Series on Nonlinear Science. World Scientific. London.

Martín-Pereda, J.A. and González-Marcos, A. (1992) Optical Programmable Processing Element using Optical Fibers, *IEEE Lasers and Electro-Optics Society,* **LEOS'92**. Boston, 15-20 November.

Martín-Pereda, J.A. and González-Marcos, A.(1995a) Some Connections between Neurophysiology and Optical Computing based on the Theory of Complexity, in *Fluctuation Phenomena: Disorder and Nonlinarity* (ed. J. Bishop & L. Vázquez), World Scientific Press, Singapur.

Martín-Pereda J.A. and González-Marcos A. (1995b) Digital chaos analysis in optical logic structures. *SPIE´s Photonics East´95 Symposium.* Philadelphia, 22-26.

Martín-Pereda J.A., González-Marcos A. and Sánchez-Guillén C. (1996) Synchronizing Chaotic Optically-Programmable Digital Circuits. *Globecom 96.* IEEE Global Telecommunications Conference, Londres.

Miller, David A. B., Daniel S. Chemla, Theodore C. Damen, Thomas H. Wood, Charles A. Burrus Jr, Arthur C. Gossard and William Wiegmann, (1985) The Quantum Well Self- Electrooptic Effect Device: Optoelectronic Bistability and Oscillation, and Self-Linearized Modulation, *IEEE J. of Quantum Electronics*, **QE-21**, No. 9, 1462-1476.

Neyer A. and Voges E. (1982) Dynamics of Electrooptic Bistable Devices with Delayed Feedback. *IEEE J. Quantum Electronics,* **QE-18**, 2009-2015.

Okada M. and Takizawa K. (1981) Instability of an electrooptic bistable device with a delayed feedback. *IEEE J. Quantum Electronics,* **QE-17**, 2135-2140.

Pecora, L.M. and Carroll, T.L. (1990) Synchronization in Chaotic Systems, *Physical Review Letters*, **64**, 821.

Pecora, L.M. (1993) Overview of Chaos and Communications Research, in "*Chaos in Communications*", SPIE Proceedings **2038**, 2-25. Bellingham, WA.

7 BIOGRAPHY

A. GONZÁLEZ-MARCOS She got her Ph.D. from ETS. Ingenieros de Telecomunicacion, (U.P.M.) Madrid Politechnical University, in 1993. She was working previously for some private companies in CAD-CAE topics. After her Doctoral Thesis, she was Assistant Professor at the Alcalá de Henares University and Associate Professor at the Alfonso X University, both in Madrid. She is presently Professor, from 1996, at the Technical Higher School of Telecommunication Engineers, U.P.M. Her present interest is on Optical Computing, on the applications of Chaos Theory to Optical Communications and on the modelling of the mammalian retina. She has published several papers on the above topics in technical journals. She is member of several technical associations.

J. A. MARTIN-PEREDA He got his Ph.D. from ETS. Ingenieros de Telecomunicación in 1971. He is full Professor at the Madrid Politechnical University from 1975 and member of several Technical Programme Conferences, mainly in Optical Communications (ECOC, OFC, IOOC) since 1985. Member of the ESPRIT Advisory Board between 1990 and 1993. Member of the Royal Spanish Academy of Engineering since 1995. Senior Member of IEEE and Member of OSA and SPIE. He has published more than 100 papers in journals and books and has got several academic and technical prizes. His present interest is on the application to Chaos Theory to Optical Communications and Biological Systems.

INDEX OF CONTRIBUTORS